How Transformative Innovations Shaped the Rise of Nations

How Transformative Innovations Shaped the Rise of Nations

From Ancient Rome to Modern America

GERARD J. TELLIS AND STAV ROSENZWEIG

ANTHEM PRESS

Anthem Press
An imprint of Wimbledon Publishing Company
www.anthempress.com

This edition first published in UK and USA 2018
by ANTHEM PRESS
75–76 Blackfriars Road, London SE1 8HA, UK
or PO Box 9779, London SW19 7ZG, UK
and
244 Madison Ave #116, New York, NY 10016, USA

British Library Cataloguing-in-Publication Data
A catalogue record for this book is available from the British Library.

ISBN-13: 978-1-78308-732-7 (Hbk)
ISBN-10: 1-78308-732-3 (Hbk)

This title is also available as an e-book.

To Cheryl Tellis
For her constant support and encouragement

To Eyal, Oren and Tomer
For their endless love and support

To Don Murray
For his belief in and generous support of our research

Contents

Illustrations

Figures

Tables

Acknowledgments

The effort for this research spanned ten years and involved an in-depth analysis of the history of innovations that transformed the fate of global powers as they peaked and declined. This research would have been impossible without the help and support of numerous people.

First, we thank Don Murray for his belief in our research and his generous support of it through grants to the USC Marshall Center for Global Innovation.

Second, a number of research assistants helped with various chapters and in collecting data, figures and graphs. We are very grateful to Daniel Charnoff, Lauren Alderette, Jeanne Almeida, Darryn Azevedo, Leslie Chang, Selin Erguncu, Rahul Francis, Kristen Mascarenhas, Sajeev Nair, Monica Orlinschi, Aviad Moreno, Connor O'Shea and especially Leila Feldman.

Third, we thank our families for their inspiration, constant support and sacrifice of time. We are grateful to Cheryl Tellis and to Eyal, Oren and Tomer Rosenzweig Soffer.

1

Global Influence of
Transformative Innovation

In the three centuries BCE, several states flourished around the Mediterranean: Athens, Macedonia, Egypt, Carthage, Phoenicia and Rome. Yet only one of them, Rome, was able to create and efficiently manage a vast empire, for over 300 years, eventually extending from England to Persia and surrounding the whole Mediterranean (see Chapter 2). What caused Rome to become such a vast, encompassing and stable empire for so many centuries?

In the thirteenth century, a self-made leader from Mongolia, Genghis Khan, united disparate tribes and built an empire that stretched from Eastern China across Asia to Russia, Hungary and the Eastern Mediterranean (see Chapter 3). How did people from a relatively poor and technologically backward country of Mongolia come to overpower and rule so many rich, established kingdoms all over Asia and Eastern Europe?

In the early fifteenth century, China was the world's greatest navigational power. At its peak, its navy, under Admiral Zheng He, had over 300 ships, which were up to 400 feet long, with a crew of 37,000.[1] It made multiple expeditions to the Far East, India, Middle East and Africa (see Chapter 4). Yet, when still at its peak, China abruptly stopped these expeditions, closed in on itself and fell behind. Many nations surpassed China as navigational powers. What caused China to fall behind these other nations?

In the fifteenth century, the city-state of Venice amassed great wealth and became the dominant power in the Mediterranean and probably the richest political entity in Europe. It monopolized trade between Europe and the East, especially the spice trade. In contrast, Portugal was a small, relatively backward country. Yet by the end of the fifteenth century, Portugal, and not Venice, discovered the sea route to India and the East that triggered the rise of the Portuguese empire and the eclipse of Venice as a naval and trading power (see Chapter 6). What led to the rise of Venice and subsequently that of Portugal?

In the seventeenth century, the relatively poor lowlands that now constitute the Netherlands united to form a great empire that at its peak controlled territories in the Americas, Africa, South Asia and East Asia (see Chapter 8). What led to the rise of the Netherlands?

In the sixteenth century, the lands that now constitute Germany were four times the size of England in terms of gross domestic product (GDP). Germany was ahead of England on most metrics of economic activity. It enjoyed a larger land mass, more forests, more arable land and more minerals (especially coal and iron) than England. Yet, by 1820, England's GDP was one and a half times as large as Germany's (see Chapter 10).[2] What caused England to speed ahead of Germany?

In the early nineteenth century, Brazil, Mexico and the United States were similar nations with abundant resources and sparse populations. In 1800, the per capita GDP of the United States was about the same as that of Brazil and only twice that of Mexico. By 1913, the per capita GDP of the United States grew to be about four times that of Mexico and about six times that of Brazil.[3] The United States had become a far bigger economy than either Brazil or Mexico and a major world power. By 1900, it even surpassed England in GDP (see Chapter 11). What caused this dramatic divergence in the economies of the United States, Brazil and Mexico?

The above cases show that economic and global leadership in the world has shifted from nation to nation. No one country has persistently led the world in terms of innovation, wealth and power. It is fascinating to see how just in the last 500 years, leadership in these areas has swung from Mongolia to China to Venice to Portugal to the Netherlands to England and to the United States. The Ottomans from Turkey and the Mughals in India also became dominant powers in between. Yet, no nation held sway permanently. Why did this constant change in global leadership occur?

Scholars in various disciplines have provided numerous explanations for the rise of great nations, as in the cases above. These explanations include advantageous geography, natural resources, climate, religion (especially Christianity), Western culture, colonization or luck. Many of these explanations have been deep, scholarly studies of one or some of the above cases. However, a robust explanation is one that can cover all the above cases, preferably with a single explanation. Most of the prior explanations, while good for one or a couple of cases, fail when applied to all of the above diverse cases, as we discuss later in this chapter. In particular, the rise of economic and world powers in the above cases occurred over time, often when geography, climate and natural resources remained essentially the same. So, these three factors cannot be a sole or primary explanation. Also, across the above cases, religion and culture varied substantially. Indeed, the rise of Rome, the Mongols, China and the Ottomans occurred without the aid of Christianity. Moreover, among Christian states, some rose and fell, while others stagnated. Some were Catholic (Portugal and Venice), while others were Protestant (the Netherlands and England). Thus, neither Protestantism nor Christianity can be the primary explanation for the cross-country rise of nations. So what factors explain the rise of the nations listed above? Are these factors similar across these diverse cases?

The first thesis of this book is that economic growth, national dominance and global leadership are fueled primarily by the embrace of innovation. In particular, the rise of nations in the cases above was probably due to the embrace of what we call transformative innovations. However, other prior, parallel or supporting innovations were also adopted in an environment that was conducive to innovation. A *transformative innovation* is a product or process that uses a completely new technology as opposed to existing products and processes, provides users with substantially superior benefits over existing products and processes, and provides adopting nations with a substantially superior edge in relative wealth or power compared to non-adopting nations. By this definition, transformative innovations are relative to the time and region in which they emerge. Once they are widely adopted and become pervasive, they are no longer innovative.

The adoption of a transformative innovation spawns numerous other related or consequent innovations. It provides a competitive advantage to a nation and may propel a small backward region to world leadership in as short a time as a century. Further, the transformative innovation can sometimes itself promote the type of positive environment that leads to further innovation. Thus, embracing innovation can start a positive cycle of wealth creation, economic dominance and a positive environment for still further innovation. This positive cycle continues as long as the environment that spawned the innovation remains supportive or until another transformative innovation arises elsewhere.

What are these transformative innovations? They are concrete for the Roman Empire, swift equine warfare for the Mongols, shipbuilding for China, Venice, Portugal and the Netherlands, the patent system and the steam engine for England, and mass production for the United States (see Table 1.1). Simultaneously, innovations in gunpowder gave various rival nations a critical edge, albeit for a relatively short time, because diffusion and imitation of this technology was rapid. Summaries in this chapter and details in subsequent chapters explain the role of these transformative innovations in the rise to dominance of select nations.

What factors characterize the conducive environment for the creation, adoption and further development of transformative innovations? The second thesis of this book is that the embrace of innovation does not occur randomly or by luck across time and nations. Rather, it is sustained by an environment characterized by key institutional drivers within a country or region. Three of the most important of these drivers are (1) *openness* to new ideas, technologies and people, especially immigrants; (2) *empowerment* of at least some individuals to innovate, start businesses, trade and keep rewards for these activities (e.g., through social status, property rights or patent laws); and (3) *competition* among nations, patrons, entrepreneurs or firms to enjoy the rewards from innovation. An explanation so simple across cases that are so diverse may appear sweeping or hasty. It is best to consider it as opening a debate about a general explanation for the role of innovation in the shifts of dominance among nations. Geography, resources, climate,

Table 1.1 Transformative innovations and the rise of nations

Year	Nation	Transformative Innovation	Prior Technology	Result
100 BCE+	Rome	Concrete	Stone, brick	Roman Empire
1200+ CE	Mongolia	Swift equine warfare	Armored cavalry	Mongolian empire
1400+	China	Water navigation technologies and instruments		Chinese dominance in trade with SE Asia, India, Arabia, Africa
1400	Venice	Galley, Arsenal		Venetian empire
1500+	Portugal	Caravel		Portuguese empire
1600+	Netherlands	*Fluyt*, sawmill	Caravel	Dutch empire
1700+	England	Patent system, spinning wheel, steam engine	Hand spinning, weaving, mining and labor/wind-based navigation	British Empire
1800+	USA	American mass production	1st Industrial Revolution	US Superpower

religion and colonization probably played a role. However, past treatments of the rise of nations have overemphasized the role of these other factors. Past treatments have downplayed or ignored the role of innovation and the institutional drivers that led to their development and adoption.

The term *institution* means organizational structures that arise from the policies of rulers or governments. One may well ask why we refer to these as institutional and not cultural drivers. Cultural analogues of each of these certainly exist. However, culture is a term that is relatively ambiguous and difficult to change in the short term. Institutions are relatively better defined and can be changed in the short term by a change of laws or the dictates of a ruler. We focus on the institutional drivers that can and were changed within and between these countries, sometimes in a short time. More importantly, institutions shape culture, and these institutional drivers create a culture of innovation. Here we give a brief definition and theoretical explanation of these institutional drivers. Details are in subsequent chapters as well as the examples below.

By openness we mean the embrace of new ideas, technologies and peoples. One people or nation, limited as they are by their environment and numbers, cannot come up with every idea all the time. Innovation thrives on the integration of ideas and methods across geographic or cultural boundaries. Thus, to capitalize on advances made in other regions and countries, a nation needs to be open to such ideas from other nations or regions.[4] We find that nations that are the most open to ideas and technologies, make the greatest advance in innovation, especially in transformative innovations that have huge payoffs. Less obvious than openness to ideas and technologies is openness to immigrants. We find that transformative innovations arose in nations around the same times as these nations had a big surge in immigration. We suggest that in numerous cases, expatriates stimulated, and were at least partly responsible for, innovation.

Why would immigration foster innovation? At least three reasons may account for such a causal relationship. First, immigrants typically enter a new nation either because they voluntarily left their home countries for better opportunities or they were persecuted in and expelled from their home countries. In either case, immigrants arrive in their new country at a great disadvantage, if not in abject poverty. This situation leads immigrants to be risk seeking and entrepreneurial because the disinheritance and poverty that accompanies immigration leaves them with few other options. The saying "necessity is the mother of invention," applies very well to immigrants. Second, immigrants bring knowledge or expertise accumulated in their prior home country. For example, Muslims and Jews who immigrated to Portugal in the fourteenth century brought expertise about navigation, commerce and finance that supported Portuguese innovations in navigation (see Chapter 7). Likewise, immigrants who fled Catholic countries for the Netherlands in the seventeenth century helped the Dutch innovations in trade, commerce and navigation (see Chapter 8). Third, immigrants facilitated the mix of cultures, people and environments that leads to cross-fertilization and innovation. Other factors that help immigrants become entrepreneurial and innovative are resilience through hardship, a better reading of social cues, and networks of other immigrants from the same country of origin.[5] Along these lines, subsequent chapters show that openness to ideas and to immigrants is an important driver of innovation.

By empowerment, we mean the granting of rights by rulers to the ruled or at least to some of the ruled. Chief among these is the right to innovate, market one's innovation to others and keep the profits that accrue from the adoption of the innovation by others. Developing innovations involve costs and risks. Marketing innovations involve even more costs and risks. Thus, innovators are unlikely to incur all these costs and undertake all these risks unless there is some incentive at the end of the journey. Keeping the profits from innovations is such an incentive. The patent system is a modern formalization of the rights of innovators to profits. The development of the patent system is itself a major innovation, and is detailed in Chapter 9. However, financial profit is only one form of reward from innovation that has become pervasive in the last couple of centuries. Other incentives for innovation include gaining social status and power, promotion through the ranks of the military or bureaucracy, or recognition by those in power with titles and public honors.

By competition we mean rivalry among nations (macro-competition) or among suppliers within a nation (micro-competition) to produce the innovation and profit from it. A great example of macro-competition is the development of gunpowder (see Chapter 4), which initially emerged in China early in the tenth century. Intense rivalry among nations in China and Eastern Asia was responsible for its early development until about the mid-fifteenth century. After that, China entered a period of relative stability with far less cross-national wars than in the prior 400 years. At about the same time, gunpowder reached Europe, probably through the exploits of the Mongols. While China experienced relative stability, Europe was immersed in intense warfare (see Figure 4.3 in Chapter 4). That intense competition led to numerous innovations in the development of gunpowder, each of which gave the nation that developed it a short-term advantage. In Asia and Europe, gunpowder did not lead to a long-term advantage, because innovations by one nation were quickly imitated by other nations.

Micro-competition is a natural consequence of the granting of rights to people or at least to certain individuals to innovate and profit from the innovation. If many or all individuals in a country have such rights, then the natural outcome is competition among these individuals to exploit the advantages from those rights. Why would rulers not allow competition? That could occur for at least two reasons. First, elites who gained power or profited from past innovations may have vested interests in the old and oppose new innovations by outsiders. Second, rulers themselves might benefit from large established organizations that grew from past innovations, through taxes, bribes or kickbacks. Thus, rulers could be opposed to new innovations that detract from the old. So, empowerment of individuals is not complete unless and until they also have the right to compete with each other or with established organizations that may supply consumers with goods that serve the same need as the innovation. A good example of micro-competition is that which existed in the United States in the nineteenth century, in contrast to the lack of such competition in Mexico and Brazil (see later in this chapter and in Chapter 11).

This book provides evidence about the role of transformative innovations in the dominance of nations and of the institutional drivers in their adoption and development. The depth of evidence varies with the age of the nation or the case under study. While the evidence may not be complete in every case, the pattern is suggestive. The strength of the argument lies in the description of a quasi-experimental situation, where two or more nations had similar starting points with similar opportunities yet vastly differing endpoints. For example, subsequent chapters discuss the rise of Rome versus other Mediterranean city-states, Venice versus Genoa, Portugal versus Venice, the Netherlands versus Portugal, England versus the Netherlands or Germany, and the United States versus Mexico and Brazil (see Chapters 2 to 11).

The natural question that pops up is, what causes the difference? Elaboration of each case attempts to show that only one of these nations (e.g., the United States and not Mexico or Brazil) developed or adopted an innovation (e.g., mass production), which resulted in the huge differences in end points. The next question that pops up is, what factors led to the development and adoption of the innovation? Case elaboration again shows that the nation that embraced the innovation had relatively more institutional openness, empowerment and competition than the one that did not. So, the strength of the argument depends on explaining with adequate evidence why initially similar nations end up with substantially differing end points, in a well-defined period.

Let's apply this logic to the cases above.

Concrete: Rome versus Mediterranean Rivals

One of the marvels of the Roman Empire is that it maintained control of all lands around the Mediterranean and those in Europe and the Middle East from England to Persia. This huge land mass posed a challenge to conquer, subdue and administer efficiently for several hundred years. How could the Romans do it? In contrast, all the other Mediterranean empires that immediately preceded it remained relatively regional or short lived. Alexander the Great's empire splintered with his death into four smaller satraps. The Phoenicians and Carthage remained relatively local. And the Greek city-states had relatively regional geographic domains.

The development and use of concrete was the transformative innovation that greatly facilitated the establishment and administration of the Roman Empire (see Chapter 2 for details). Roman concrete was made from pozzolana that was mined from volcanic ash. It was mixed with lime, sand, water and aggregate to make types of concrete of varying strength. Roman concrete was used to keep stones and bricks together. But it was also poured to fill the core of a wall or pillar that was formed from brick facings or wood form. A type of concrete that could harden under water was used to build ports.

Roman concrete was used for roads, ports, aqueducts, walls, theaters, amphitheaters, fora, temples, palaces, markets and houses. It enabled massive construction of these structures, some of which remain until today. despite destruction and mining in subsequent centuries. The roads and ports enabled travel, trade, conquest and administration. The other structures all served important aspects of Roman life including worship, entertainment, water distribution, trading, bathing and living.

Rome definitely had a geographical advantage in possessing one of the raw materials for concrete in the many volcanic mines around the city. But Greeks, Egyptians and other nations around the Mediterranean also mined volcanic ash and developed forms of concrete. However, the Egyptians built with stone blocks cut to shape and joined by cement. The Romans developed a form of concrete that could be poured or dropped into forms set by brick or wood. Building by pouring concrete was quicker, cheaper, stronger and required fewer skills than building by chipping stone and laying one stone or brick upon another. Roman concrete became the important means of building and made obsolete solely stone or brick construction.

What led to this innovation in Rome? Chapter 2 details how an environment of relative institutional empowerment, competition and openness in the Roman republic and early empire spurred the development and extensive use of concrete. Romans were open to embracing and building on the architecture, engineering, science, art, sculpture and other forms of knowledge from the states they conquered. They were open to non-Romans participating in the work and the building of the empire. Indeed, the empire established a system of citizenship that empowered individuals of different nationalities with rights and opportunities unrivaled for its time. Citizens, though of different categories, were free to travel around the empire, work, own land and rise through the ranks of the army and bureaucracy. There was intense competition among Roman leaders to build better and more extravagant structures than their rivals and predecessors. All of these factors combined to provide numerous Roman innovations in concrete, engineering and construction (see Chapter 2). These innovations played a critical role in the stability, flourishing, wealth and durability of the Roman Empire. Concrete not only spurred these innovations but also provided a strong foundation for advances in culture, architecture and the rule of law. Indeed, these outlasted the Roman Empire and meaningfully impacted future generations.

Swift Equine Warfare: Mongolia versus the East and West

How did the scarcely populated and poor country of Mongolia come to rule a great part of Asia and Eastern Europe in the thirteenth century? Most narratives of the rise of the Mongols attribute it to the leadership or brutality of Genghis Khan. He may have been an ambitious, shrewd and ruthless leader, but at the core of his success was a major innovation in warfare. The Mongols transformed a natural resource, the Mongolian horse, into a swift war machine that neither Asia nor Europe had seen or could defend against (see Chapter 3).

The Mongolian horse breeds abundantly in the wild in Mongolia even today. It is relatively small and light with great endurance, able to gallop long distances without stop. From an early age, Mongols learn to capture wild horses and tame them for food, work, racing, hunting and warfare. They use the milk as food, and fermented milk as an alcoholic drink. They mastered shooting arrows with both hands while riding.

Since Roman times, Western and Asian nations had been resorting to the increasing use of heavy defensive armor for men and horses. This buildup of armor prompted the need for heavier and sturdier horses. Thus, over several hundred years outside Mongolia, horses were bred to be stronger and sturdier. A side effect of this buildup was that cavalries grew to be slower and less maneuverable. Genghis Kahn's innovation was to use the

Mongolian horse for swift or stealth attacks that exploited the weaknesses of his enemies' heavy and slow defenses.

Genghis Kahn and his lieutenants used Mongolian horses to develop a swift, highly maneuverable fighting machine that we term swift equine warfare. Mongol warriors typically traveled with three to five horses each. This ratio enabled them to maintain an overall rapid speed of movement each day, as riders could switch to fresh horses when their first horse tired. The army traveled through river valleys and grasslands, so the horses had natural resources to feed and drink. Thus, there was less need to carry food and water for them. Soldiers used horse milk and meat as food. They also bled their horses for nourishment and drink when they had no other source for the same. Thus, they could travel with minimal provisions of food and water, for man or animal. As such, the Mongolian horse formed the core of a fighting force that was frugal, easily built up, relatively light and highly maneuverable.

In addition, Genghis Kahn used a variety of fighting techniques that exploited these advantages of the Mongolian horse. He avoided participating in frontal pitched battles with the enemy or engaging in long sieges. Instead, he would make lightning raids on unsuspecting communities, then quickly withdraw. Alternatively, he would approach a village stealthily posing as a trader and attack when defenses were down. When faced with a large army, he would draw the enemy out from its base by feigning a retreat. Then, when the enemy was stretched out over long distances and tired, he would ambush the enemy from the sides where they were most vulnerable. When faced with heavily fortified cities, he would attack surrounding villagers and send their people into the fortified city to strain resources and create fear and confusion. He would trick and threaten enemy cities into surrendering rather than fight costly battles. For example, he would light multiple fires by night and kick up massive plumes of dust by day while riding to give the enemy the image of a much larger force than he actually had. With each victory, Genghis Kahn increased his wealth from captured towns and increased manpower for his army from captured peoples. He won over the people by eliminating the ruling class and granting people land and rights.

By the time of his death in 1227, Genghis Khan's empire stretched from the Pacific Ocean to the Caspian Sea including northern China, Afghanistan and parts of Russia (see Chapter 3). By the time of the death of his grandson, Kublai Khan, the empire expanded to include most of China, Persia, Iran and the Baltic states. It covered a vast area of almost 13 million miles that was larger than the Roman Empire or all of North America. For a single nation of about a million nomads to rule such a vast empire required great administrative skills.

How did the Mongols harness the talent and resources to build and administer such a huge empire? Chapter 3 details how the Mongols did so by empowering the people, establishing a rule of law, tolerating religious differences, discouraging tyranny and leader cults, and encouraging trade, migration and cultural exchange within and across the empire. At the empire's peak, the Mongols welcomed and encouraged debate among leaders of all major religions. They attracted engineers and technicians from various parts of the empire and moved them to areas of greatest need. The Silk Route flourished under the Mongols, and Eastern ideas and goods reached Europe, most likely stimulating the Renaissance (see Chapter 3). The Mongols quickly embraced gunpowder technology when they encountered it in China, and probably led to its dissemination to Europe (see Chapter 4). While the Mongols earned a reputation for brutality, Chapters 3 and 4 detail how they embraced and advanced innovations through great openness, the empowerment of conquered people and the fostering of competition among those they ruled.

Shipping: China, Venice, Portugal, Netherlands

Why did China become the world's greatest naval power in the early fourteenth century? Why was Venice perhaps the wealthiest nation in Europe in the fifteenth century? Why did Portugal surpass both these countries in naval power in the sixteenth century? Why did the Netherlands surpass Portugal? Transformative innovations primarily in oceanic navigation and shipbuilding are responsible for all four of these major transitions in global power and ensuing wealth (see Chapters 4 to 7).

Under the emperor Zhu Di in the early fourteenth century, China developed many innovations that enabled it to build a massive navy and master oceanic navigation. Chinese innovations included multiple water-tight compartments in ships, movable sales, the sternpost rudder and the compass. Chinese ships dwarfed those in which Christopher Columbus sailed to the Americas 60 years later. The naval force that China developed under Admiral Zheng He enabled seven expeditions across most of Asia, from Japan to southwest India, the south coast of Arabia and the west coast of Africa. China was the unchallenged ruler of the Asian seas and could depose rulers in a variety of Asian lands or require them to show up at Chinese court to pay homage or join in celebrations (see Chapter 4).

Yet by 1440, all that was history. The successor to Emperor Zhu Di shut down the navy, grounded the ships and burned the maps. On the verge of being a great trading and colonial power, China closed its doors to the outside world and closed in on itself. Innovation ceased, growth slowed and the economy stagnated. What caused this great reversal? To begin with, Emperor Zhu Di, who championed China's oceanic navigation, died. Beneath the surface of the entrepreneurial leadership of Zhu Di lay a huge bureaucracy that ran the administration. These bureaucrats were steeped in tradition and taught to conform to tradition, rather than to be novel or innovative. Their fear of losing power to the entrepreneurial explorers along the south coast of China, aided by threats from invaders in the north, prompted them to convince the new emperor to close China to the outside world, despite the great promise of oceanic navigation. The bureaucrats had little appreciation of naval power, oceanic trade, contact with the outside world or innovation because their status and power were based on tradition. China's huge stride in innovative naval technologies came to a sudden end. Lack of competition due to the dominance of a single sovereign resulted in no deviations from the new emperor's sweeping mandate about naval expeditions (see Chapter 5).

Around that time, the country that rose to prominence on navigational innovations was the Venetian Republic. In the fifteenth century, Venice was a hotbed of creativity, with innovations in business organization, shipping, shipbuilding and numerous other areas. Critically, Venice boasted the Arsenal, an impressive factory for the mass production of the Venetian galley. At its peak, the Arsenal could produce at least one galley per day. Venice won big orders from other nations for shipbuilding. With a navy of galleys, between the twelfth and fifteenth centuries, Venice colonized a large number of port cities and islands along the eastern Mediterranean, from Venice all the way to Constantinople. Through the fourth crusade, Venice stormed Constantinople and for 50 years rendered it a vassal of the Venetian empire. In trading with Western Europe, Venetian galleys went up the western coast to London in England and Bruges and Antwerp in Belgium. Venice was the preeminent European naval power and on the cusp of becoming a global one (see Chapter 6).

Chapter 6 details how Venice's innovations were facilitated by a relative openness to ideas and innovators from different parts of Europe and lands surrounding the Mediterranean. Openness was critical for Venice as a port city with a rich history of trading and as a meeting place of traders and goods from all over Europe, Africa and Asia. In addition, in the early years of the empire, Venice passed laws that empowered new entrants to easily enter the trading class, take risks and pursue fortunes that they could keep. As a result, competition among traders and builders of ships or workers in the Arsenal was intense and productive (see Chapter 6).

Yet, in the late fifteenth century, a small, relatively poor country trumped Venice not by war but by innovation. During the fifteenth century, Portugal made steady advances in navigational innovations for cross-Atlantic travel. On that knowledge, Vasco da Gama swung around the Cape of Good Hope in the southern tip of Africa and reached the shores of southern India in 1498. He opened a sea route to India and Asia. This innovative leap opened a windfall for Portugal and many other Western European countries that piggybacked on Portugal's breakthrough (see Chapter 7). However, it bore painful losses for Venice, the Ottoman Empire, India and Southeast Asia, which relied on the traditional trade routes. What was Portugal's key innovation? It was the caravel that made Venice's galley obsolete.

The caravel was a relatively small ship, light with square and lateen sails. The design of the sails made it fast and highly maneuverable. It was well suited for traversing the Atlantic Ocean, which was beset with strong currents, winds and shoals. In contrast, the galley, which Venice perfected to traverse the

Mediterranean Sea and the English Channel, was less suitable for such journeys. The relatively light displacement tonnage of the caravel also enabled it to go upstream in rivers that flowed into the ocean. The caravel revolutionized oceanic shipping. By discovering the sea route to India and the East, Portugal broke the monopoly that Venice had in trade with the orient in spices, silk, jewelry and cotton. The humble caravel rendered the Venetian galley obsolete (see Chapter 7).

Why did Portugal develop the caravel? Chapter 7 details how openness to ideas and immigrants, empowerment and competition within the country and against neighboring powers played an important role. The Portuguese had faced intense competition from raiders from North Africa, who would harass and loot villages that lay along the coast. The caravel emerged partly in response to this competitive challenge for survival and safety. In particular, Henry the Navigator was a leader who valued oceanic navigation and championed the caravel. He obtained resources from the emperor, which he used to underwrite the development of the caravel and expeditions in the Southern Atlantic and Indian oceans. At the time, the country was relatively open, borrowing extensively from Turks, North Africans, Arabs and others who had knowledge about navigating the oceans. Moreover, during the development of the caravel in the 1400s, the country was relatively open to immigrants including Muslims and Jews and benefited from their expertise in navigation, ship construction and trade.

The caravel declined with the rise of a major Dutch innovation: the *fluyt* (see Chapter 8). Initially Dutch merchants used *karveels*, the equivalent of the Portuguese caravel. By 1550, the *karveels* were replaced by smaller, more maneuverable ships called *vlieboot* and *boeier*. But rapidly growing maritime trade required larger, faster and more economical ships. This need led to the innovation of the *fluyt* in the 1590s. The innovation was in the design of the *fluyt*. Drawing from their expertise in exporting large amounts of fish and grains, the Dutch designed the *fluyt* with an extremely large cargo space, a flat bottom and a shallow draught (in which the distance between the waterline and keel—the centerline at the bottom of a ship's hull upon which the rest of the hull is built—was very small). It could carry about 360 tons. This design enabled the *fluyt* to navigate the shallow waters of ports and rivers better than ships of other nations. The *fluyt* was built of both pine and oak. The oak gave it strength, while the inclusion of pine made it very light. In addition to these features, the ship was steady in bad weather relative to other ones.

Another important Dutch innovation that contributed to its maritime power was the development of the wind-driven sawmill. Probably developed in 1594 by Cornelis Corneliszoon, the mill allowed for large amounts of wood to be sawed faster and with substantially less labor than any other state—including the Portuguese and the English.[6] In the 1670s, the Dutch fleet was bigger than the combined fleets of England, France, Spain, Portugal and Germany.

These innovations were driven by the novel institutional environment in the Netherlands that was quite unique in Europe (see Chapter 8). In the seventeenth century, the Netherlands was one of the most open countries in the world, welcoming immigrants from other parts of Europe seeking work or fleeing religious sanctions or persecution. Laws were also relatively liberal and provided individuals with property rights, the freedom to trade and the right to keep profits from innovations. This environment spawned a large number of entrepreneurs, who competed for profit. With these and related innovations, the Dutch became a great seafaring nation and built a world empire with possessions in Europe, America, Africa, South Asia and East Asia (see Chapter 8). This empire was dominant in the global sphere until it was surpassed by the British Empire.

Patenting: England versus Germany

In 1500, the English economy was less than a third the size of Germany's. By 1820, the English economy had grown to one and a half times that of Germany. During these three centuries, in terms of GDP, Germany tripled in economic size. England grew 11 times as large. Why did England race ahead of Germany in the eighteenth century? The development and adoption of a patent system played a critical role (see Chapter 9). The patent system spawned a series of innovations, most importantly, the steam engine. This innovation transformed the manufacture of good for consumers (see Chapter 10).

A patent for a design or product grants the inventors of that design the right to solely produce and sell that product for a limited time without imitators who use the same design. In effect, it is a grant of a limited monopoly for an inventor of a design. It encourages people to innovate, benefit from their innovation and disclose their invention without fear of infringement. While the monopoly leads to higher prices, the ensuing excess profits serve as motivation for innovators to invest in research and marketing and to develop and commercialize an innovation. At the same time, the disclosure of the design encourages others to invent around the patent and compete away some of the excess profits. So ironically, even though a patent provides a limited monopoly, it inherently encourages competition and further innovation.

For most of history, prior to the fifteenth century, monarchs and political elites owned landed property that was then leased to renters or serfs for farming. Similarly, the concept of intellectual property rights was either non-existent or present in only arbitrary or primitive forms. One such primitive form was prevalent in fourteenth century England. The monarch or regional leader (e.g., a duke) would give a letter of protection granting some rights to an individual or group to practice a profession or make and sell a product. These rights were granted on a case-by-case basis rather arbitrarily, at the pleasure of a political leader. However, these grants created local monopolies, granting rights to some at the expense of preexisting groups and traders. Starting in 1624, as part of a power struggle between the English monarch and the House of Commons, the latter attempted to establish a fairer system of patenting. After several attempts, the House of Commons, in 1624, nullified all prior patents and established a new Statute of Monopolies, which formalized the granting and operation of patents. Henceforth, patents became a right of the people rather than the privilege of a political leader to grant. As such, it was an enormous innovation in its own right.

Patenting led to a surge in innovations in England in the cotton, power generation and mining industries, among others. These innovations started what has been called the Industrial Revolution in England. As a result, England grew to be much more innovative, wealthy and powerful than Germany. The

English innovations were copied in other countries. Even then, the copying took time. The innovations led England to leap ahead of larger rivals in Europe and Asia.

Why did the patent system take off in England and not in Germany or other countries of Europe? Chapter 9 details how it did so, because England had a greater level of relative openness, empowerment and competition. First, around the seventeenth and eighteenth centuries, England became the most open country in Europe to new people and ideas, even more than the Netherlands. While religious persecutions were common in northern and southern Europe, England was relatively tolerant. As a result, scientists, philosophers and thinkers flocked to England to practice their profession more openly than in their original homeland. The patent system attracted innovators and entrepreneurs who wanted to profit from the limited monopoly that arose from patents. Second, since its establishment, the House of Commons was constantly in a struggle to wrest power from the monarch and empower the people. England also had a higher level of education than most other countries of Europe, empowering people with knowledge, understanding and freedom of movement. Third, England had the most intense competition of any other European country. These factors led England to slowly develop a good patent system, which triggered many subsequent innovations especially in weaving, mining and the steam engine (see Chapter 10). England surged ahead of Germany and all other European countries.

Mass Production: United States versus Brazil and Mexico

Why did the United States dramatically pull ahead of Brazil and Mexico in the nineteenth century? It did so due to the adoption of a system of mass production in all walks of life (see Chapter 11). We use the term "mass production" here to refer to the manufacturing, distribution, marketing and purchase of standardized products on a massive scale with increasing automation and decreasing labor. The hallmarks of mass production are standardization, automation and scale. One important result of mass production is decreasing variance from human errors and increasing quality, as machines can produce to tighter specifications than humans. A second result of mass production is speed, as machines can produce in less time than humans. A third result of mass production is efficiency, because in many situations machines costs less than human labor. Mass production in a primitive form and in limited time was available in ancient China. Venice used mass production in shipbuilding, at the Arsenal (see Chapter 6). England used some mass production for textiles and mining with the onset of the Industrial Revolution (see Chapter 10).

However, throughout the nineteenth century and thereafter, the United States dramatically increased the ratio of machines to labor to a whole new level. Mass production permeated many industries, from agriculture and husbandry, to the making of soap, shoes and clothing, to the manufacture of guns, tools and appliances, to the building of ships and trains. The level of mass production gradually exceeded not only that in Brazil and Mexico but also even that in the advanced European nations, which at that time carried out production traditionally with a heavy labor component. This system of pervasive mass production came to be called the American system. So, this book treats mass production as an innovation in its own right. The thrust to mass produce goods itself led to numerous innovations in various industries. In so doing, within a hundred years, the United States surpassed Brazil and Mexico, and equaled or surpassed England and other European nations in the quantity of manufactured goods, if not also their quality. As such, US imports of manufactured goods dropped substantially, while US exports of manufactured goods relative to raw materials increased substantially. It became a major force in the world economy (see Chapter 11).

One driving force for the adoption of mass manufacturing was the abundance of land coupled with scarcity of labor force. For example, during the nineteenth century, westward expansion of the United States led to the acquisition of numerous states including Louisiana and the Midwestern states and most of the states west of the Mississippi River. However, in terms of land, Mexico and especially Brazil were almost equal to the United States. Brazil had a landmass not much smaller than that of the United States. Why did these two countries not embrace the innovations that flourished in the United States or at least copy them? The reason is that the environment in Brazil and Mexico was dramatically different from that in the United States. Briefly, the United States was far more open to new peoples, empowered people with more rights and opportunities, and encouraged more intense competition than did either Mexico or Brazil. Let us consider each of these factors briefly here. Details are in Chapter 11.

First, the United States welcomed immigrants from many countries and from all classes of people in each country. In contrast, Brazil and Mexico welcomed people only from the parent colonial powers (Portugal or Spain, respectively) and within those countries, only elites. As a result, the population in the United States exploded from 5 million to 76 million during the nineteenth century. In contrast, during the same time interval, the population increased in Mexico from about 5 million to 13 million, and in Brazil from about 3 million to 17 million (see Chapter 11). Moreover, the arrival of so many new peoples from a variety of countries created a mix of ideas, styles and cultures that produced a fertile breeding ground for innovation.

Second, the United States empowered its citizens as few countries at the time did, and much more so than Brazil or Mexico. This spirit of empowerment was rooted in the US Declaration of the Independence and the Constitution. The United States was founded on the principle of equal rights. In various periods, equal rights were not given to certain groups (slaves, women and immigrants from China to name a few). Still, this equality intended to include men of most countries of origin and ethnicities and all religions and classes. In contrast, Brazil and Mexico enacted preferences to people from select countries (Portugal and Spain, respectively) and select religions (Catholics). In the United States, new entrants were free to own land or start businesses. However, in Brazil and Mexico, land and business were reserved for elites. One law epitomized empowerment in the United States. Beginning in 1962, the US Congress passed the Homestead Act, which granted large tracts of land to any adult man or woman meeting certain minimum standards, if they went west to cultivate the land. Subsequent amendments expanded the act in various ways. This act spawned a great movement westward of peoples to launch out as entrepreneur farmers. In contrast, Mexico and

Brazil had a system of huge farms run by vested elite farmers who used laborers as serfs and were interested in sustaining the status quo (see Chapter 11).

Third, the United States fostered easy entry into markets by new entrepreneurs and fair competition among farmers and manufacturers. In so doing, it fostered intense competition in markets for producing farm and manufactured goods. The above policies of openness and empowerment created a class of millions of entrepreneurs who were eager to take advantage of the opportunities in the new land and create personal wealth and success in one or another market. This intense competition in turn triggered extensive, constant innovation in small and big ways. Innovation was imperative to survive and succeed against rivals with same opportunities and resources. In particular, a huge drive ensued to manufacture both farm and nonfarm products with maximum automation and minimum labor, on a massive scale, at low cost and high quality. Thus was born the phenomenon of mass production in the United States (see Chapter 11). The drive for mass production led to innovations in every industry and market, spanning agriculture, soap, shoes, clothing, guns, tools, appliances, shipping and trains. In contrast, Brazil and Mexico had far more restrictions on competition. They both had large estate farms, with prescribed markets, few rivals and much less competition than the United States. As a result, Brazil and Mexico had a much lower drive for automation and large-scale manufacturing with little to no innovation. In comparison to the United States, Mexico's economy grew slowly, while Brazil's economy even shrank.

Figure 1.1 presents a model of the theory proposed in this book. It shows how three institutional factors drive transformative innovation, which in turn leads to national wealth and dominance. The wealth and dominance are likely to spur further innovation, and even further transformative innovation. Further innovation has the potential to generate a positive cycle that further feeds openness, empowerment and competition. However, wealth and dominance may also lead to negative consequences of focusing excessively on past success and innovations, and not venturing into the next transformative innovation. Indeed, long-term success and dominance, which can be elusive, depend on countering the curse of success by ensuring continual openness, empowerment and competition. Thus, the positive cycle continues either until openness, empowerment and competition decline, or until another transformative innovation arises elsewhere.

What about Other Explanations?

Authors have proposed various other explanations for the rise or dominance of nations, including geography, natural resources, climate, Christianity, Western culture, colonization, luck or technological cycles. Let us briefly review these explanations against the backdrop we have just sketched.

Probably the best case for geography was made by Jared Diamond.[7] Diamond argued that certain civilizations grew because of their geographic location that provided a competitive advantage over others. This advantage arose from the availability of wild animals, seeds and plants that could be easily domesticated. Such domestication in turn led to further innovations in selective breeding, crop raising, food storage, transportation and related spheres of life. Civilizations that lay along the contiguous geographical and climate belt of Euro-Asia enjoyed trade and the diffusion of the best-performing plants, animals, related innovations and innovation spillover. These civilizations flourished, advanced, spread and gained dominance. As a byproduct, these people developed infectious diseases from the animals they bred. But long exposure to such diseases over millennia lent them immunity, which those in other noncontiguous continents did not have. Successful agriculture, coupled with the easy availability of minerals, wood and coal, led nations to develop steel and guns, which gave them a competitive edge over other nations that did not have these resources. Diamond's thought-provoking and masterful thesis arguably provides a powerful explanation for why Europeans colonized the Americas, Australia and sub-Saharan Africa and not vice versa.

However, when one focuses on the dominance of nations within the Euro-Asian belt, the role of geography and resources is relevant but not dominant. The reason is that geography and resources are relatively constant over time, but the dominance of nations varies over time. Time-varying innovations are a better potential explanation of the time-varying rise of nations than constant geography or resources. How so? Knowledge in the form of technology evolves over time. During that time, small incremental changes accumulate. If these changes are accepted and implemented, they ultimately cross a threshold constituting a major innovation that makes new resources valuable and renders old resources or social structures obsolete. These are transformative innovations. In contrast, geography and resources remain relatively constant. It is technological innovation that renders geography relevant. It is technological innovation that renders resources productive.

For example, between the twelfth and fifteenth centuries, several city-states had the advantage of being located on the shores of the Mediterranean with the ability to dominate it, colonize other city-states and exploit the wealth in cross-Mediterranean trade. Yet over a period of time, only one, Venice, developed the innovative Arsenal that mass produced the Venetian galley, which enabled its dominance of the Mediterranean and the monopolization of Mediterranean trade for a time. When Portugal developed the caravel, it rendered the Venetian galley and Arsenal obsolete. Likewise, Portugal always had the potential of being located by the Atlantic Ocean, across which one could navigate to reach Brazil or India. However, until the development of the innovative caravel in the fifteenth century, that potential did not amount to much. When the Dutch *fluyt* displaced the caravel, Portugal's advantage vanished (see Chapter 8). The same pattern can be observed in the rise of all the other nations mentioned in this chapter and developed in subsequent chapters. Each had some geographical or resource potential. But it was the adoption of a transformative innovation at a particular point in time that rendered that geography advantageous or that resource productive. Likewise, when another nation developed and adopted some other transformative innovation, the geographic or resource potential declined in importance.

The climate or latitude theory of economic growth argues that distance from the equator is the primary explanation for economic progress across nations.[8] A proposed reason is that warm climates create lethargy, thus discouraging work, while cool climates do the opposite. Looking at Europe in the late twentieth century, one can certainly see a gradation in economic development and progress as one moves from the warm south to the temperate mid-regions and the cool north. Indeed, one of the current author's own studies found that, in the adoption of modern consumer innovations, the Nordic countries are the most innovative, followed by the mid-Atlantic countries and, last, the Mediterranean countries.[9] Does this mean that climate is the cause for the progress of nations? The fallacy of climate can be seen as soon as one withdraws from the limited focus of the late twentieth century and looks across a sweep of centuries and civilizations, as done earlier in this chapter. The climates of Rome, Venice, Portugal, Germany and England, relative to each other, have not changed dramatically in the last 2,000 years. Yet their relative rise and decline have. Climate cannot explain these changes. The adoption of transformative innovation, as argued above and in subsequent chapters, can.

One of the most prevalent theories of the rise of nations is that of Western culture in general and Christianity in particular. Besides Max Weber's classic on the Protestant work ethic, recent articulations of this position in various forms have been espoused by Ian Morris (*Why the West Rules—For Now*); Rodney Stark (*How the West Won* and *How Christianity Led to Freedom, Capitalism, and Western Success*); David L. Landes (*The Wealth and Poverty of Nations*); and Joel Mokyr (*A Culture of Growth*).[10] It is challenging to do justice to these great treatises in a short space. We submit that if one defines Western culture as the norms and values of Western Europe and restricts the time period from about 1500 to 1900, then the data may seem consistent with the theory. Highly innovative and rapidly growing nations in this time period had Western values and were Christian states. If one considers the Americas as inheritors of Western values and Christianity, then the argument holds even to the middle of the twentieth century.[11]

However, the argument begins to fail when one considers the rise of Japan in the second half of the twentieth century, the rise of Southeast Asia in the last 30 years and the recent rise of China. Even then, defenders of the Christian and Western culture theories may say that Japan, Southeast Asia and China succeeded when they adopted Western culture or Christian norms and values. However, these theories really fail when one considers non-Christian China in the fifteenth century, non-Christian Mongolia in the thirteenth century and pre-Christian Rome until the fourth century. Indeed, for most of the time from the fall of the Roman Empire to the end of the fifteenth century, dominant world powers were Muslim empires, Mongolia and China.[12] None was Christian, and all had cultures quite different from that of Western Europe. Indeed, prior to the fourteenth century, most of the major innovations originated from Asia, particularly China and Mongolia, including swift equine warfare (see Chapter 3), gunpowder, cannons and guns (see Chapter 4), the compass, movable sails and watertight compartment for ships (see Chapter 4), the blast furnace, paper, paper money, the printing press and many others. Theories of Western supremacy in innovation ignore almost a millennium of innovations in Asia.

The theories of Western culture or Christianity collapse especially when one considers the dramatic rise and fall of nations *within the Western world.* Christianity or Western culture just cannot explain the rise of Catholic Venice versus other Italian republics and its falling behind Portugal, the rise of Catholic Portugal and its falling behind the Protestant Netherlands, the rise of Protestant England over other European countries and its falling behind the United States, and the rise of the secular United States versus Brazil and Mexico. Transformative innovations, facilitated by openness, empowerment and competition, can, as explained in Chapters 6 to 11.

In a variation of the explanation of Western culture, Joel Mokyr argues that the world economy changed after 1700 due to the development of the steam engine, which triggered the Industrial Revolution. Knowledge development and innovation were cumulative after that watershed event but isolated before.[13] We do not subscribe to this theory. While the development of the steam engine introduced a revolution in power generation and consequent mechanization (see Chapter 10), there is no reason to believe that it marked a watershed moment in all knowledge development and innovation. Knowledge development and innovation have been cumulative even in prior periods. For example, Chapter 4 shows how knowledge about gunpowder was cumulative, tracing a continuous path of innovations from before the ninth century to the seventeenth century. During these centuries, innovations from the Chinese, Mongols, Ottomans and Western Europeans built on prior innovations and led to new innovations. Thus, knowledge and innovations were cumulative even before the modern economy. Likewise, Chapters 5 to 8 trace the history of innovations in navigation from before the fifteenth century in China through Venice, Portugal and Netherlands in the seventeenth century. An implicit yet important premise of the present book is that there is no single period in time that can be considered a watershed moment, before which innovations were isolated and after which innovations were somehow cumulative and continuous.

A variation on the thesis of Mokyr was offered by Kenneth Pomeranz, whose thesis combines a focus on the Industrial Revolution with a variation of the geography thesis.[14] Unlike some of the other scholars tackling the subject, Pomeranz is not Eurocentric. He suggests that on the eve of the Industrial Revolution there were remarkable similarities between Western Europe and the most advanced regions of Asia and certain regions of China in particular. According to Pomeranz, evident similarities included land, labor, living standards, products and markets. In other words, Western Europe was not as far ahead of China as one may think. Similarly, both regions suffered from shortage in land and timber. However, England, as opposed to China, was able to escape this shortage trap due to (1) the availability of coal and (2) accessibility to America, which offered not only a source for raw materials but also a critical market for finished goods. Pomeranz's analysis is largely limited to a comparison of inputs and outputs in China versus Western Europe and does not address the broader social and technological developments that accompanied the Industrial Revolution.[15] More importantly, his thesis attempts to explain merely the divergence between East and West that followed the Industrial Revolution, and cannot explain the rise to dominance of some nations as opposed to others in earlier periods. Not only the availability of coal or lack thereof was irrelevant prior to the Industrial Revolution, but also the availability of other energy sources did not make a difference. For example, Germany had coal, but England rose to dominance whereas Germany did not. Timber was far from abundant in the Asian steppe, but the Mongols relied on innovation—not energy—to gain dominance. Similarly, sixteenth-century Venice had access to numerous markets for finished goods, but so did its Mediterranean city-state rivals that did not achieve the prominence of Venice. England was able to take advantage of coal and of its proximity to the Americas not because of some geography-based "good luck" but because the transformative innovations of the steam engine and the patenting systems that enabled it to turn these resources into meaningful assets.

A work that relates to ours is *Why Nations Fail* by Daron Acemoglu and James Robinson. Both works emphasize the role of institutional drivers in the performance of nations.[16] Acemoglu and Robinson argue that nations that are blessed by inclusive institutions flourish, while those that have extractive institutions fail. Extractive institutions are those in which the rulers pass laws and manage the economy to benefit a few in the ruling class, while letting the majority of the population languish in abject poverty. With extractive institutions, rulers preside over a transfer of wealth from the population to themselves. Acemoglu and Robinson's thesis and ours rely on institutions as drivers of prosperity. However, three major distinctions exist between the two theses. First, Acemoglu and Robinson focus on the failure of nations, while we focus on the rise to dominance of nations, and subsequently the rise of other nations, depending on innovation. Many factors play a role in the failure of nations and empires, as subsequent chapters discuss. Second, Acemoglu and Robinson do not postulate a mechanism through which various institutions drive success or failure. In contrast, we use innovation as the central phenomenon between institutions and the creation of wealth, power and dominance (see Figure 1.1). In our view, the institutional drivers of empowerment, openness and competition lead to the development of transformative innovations, which in turn bring great wealth, power and dominance for a particular nation. Third, we focus on the dynamics between nations, in the sense that wealth, power and dominance do not guarantee further transformative innovation. Sometimes, success itself sows the seeds of failure. Excessive focus on a past transformative innovation blinds a nation to transformative innovations of the future. Another nation embraces openness, empowerment and competition, and develops the next transformative innovation,

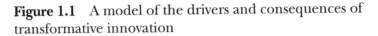

Figure 1.1 A model of the drivers and consequences of transformative innovation

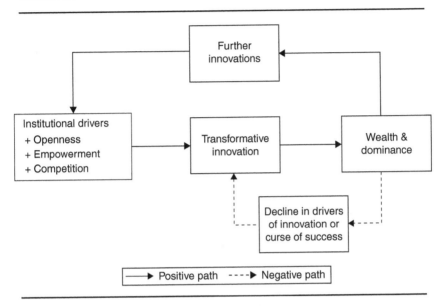

leading to its subsequent rise and the decline of the prior leader. Thus, decline comes not so much from extractive institutions, though the latter could be detrimental. Decline comes from suppressing openness, competition and empowerment, or by focusing excessively on the past and not innovating relentlessly to develop and embrace the next transformative innovation.

A case in point is the Venetian Empire in the thirteenth to fifteenth centuries (see Chapter 7). Acemoglu and Robinson suggest that the Venetian Empire began declining toward the end of the thirteenth century as the privileged class of successful merchants changed the laws to enrich themselves by preventing entry of and competition from new traders, who were typically responsible for the creation of trade and wealth in the empire. In contrast, we find that the empire continued to grow in power and wealth until the end of the fifteenth century. In 1498, the Portuguese navigator Vasco da Gama successfully rounded the Cape of Good Hope in South Africa and reached the shores of India. That trip meant that Portugal could get the spices, silks, cottons and jewelry of the East directly without going through the long and costly mediation of merchants in the Middle East, Turkey and North Africa, as Venice did. Portugal had undercut the very source of Venice's wealth by taking advantage of its own transformative innovation.

Moreover, the basis of Portugal's strength was its innovative caravel, which exploited wind power and could navigate the turbulent waters of the South Atlantic. The Portuguese had taken about a century to perfect the caravel for this task (see Chapter 7). In contrast, Venice's strength lay in the Arsenal that mass produced the galley. Venice had focused on perfecting the Arsenal, which at its peak could produce about one galley a day. The Venetian galley relied primarily on human power (oarsmen) and was good primarily for navigating the relatively stable waters of the Mediterranean, which separated Venice from its trading partners in Turkey, the Middle East and North Africa. Unlike the caravel, the galley was unfit for navigating the South Atlantic. The Portuguese caravel had obsoleted the Venetian galley (see Chapter 7). Venice's excessive focus on the galley, the Arsenal and the Mediterranean probably blinded it to the importance of the direct sea route to India and the East via the Atlantic. By that time, Portugal leaped forward with an unusual openness to ideas from abroad, and with empowerment and intense competition.

What about luck? Was not a nation (or an innovator) just lucky in stumbling upon and adopting an innovation? A third premise of the present book is that innovation is not a lucky, random or serendipitous event but the fruit of a supportive environment. That environment again is not random but manageable, arising from three institutional drivers. Luck is a particularly pernicious explanation because it relies on an unidentified "other" factor. The proponent of luck just asserts that adoption occurred because of a fortuitous coincidence of "the right time and place."[17] The luck theory of innovation has considerable support from the casual observer, who lacks knowledge of institutional detail.

At the individual level, this theory has been nicely debunked by a series of clever experiments by Richard Wiseman, who shows that creative people are opportunistic, resilient, outgoing and optimistic, and consider themselves "lucky."[18] In contrast, noncreative people lack these four attributes and also consider themselves unlucky. But by practicing the behaviors of the creative, the latter can become more creative. Thus, luck is not a cause. Certain personality traits make people creative and may consequently make them successful. Along the same vein, a meta-analysis of research in psychology shows that creative people are more open to new experience, less conventional and conscientious, and more self-confident, self-accepting, driven, ambitious, dominant and impulsive.[19] Thus, innovation at the individual level arises not from luck or happenstance.

At the firm level, the theory of luck has been debunked by some powerful treatises that show how innovation rises from the underlying characteristics of innovators such as experimentation, persistence, questioning and openness to diverse experiences,[20] or from certain economic or social conditions in the environment.[21]

Luck implies being "at the right time and place." To paraphrase, it implies that a nation rose because of a fortuitous combination of events and locations. But therein lies its fallacy. The argument of luck is essentially circular. It identifies no exogenous factors of causation. It merely labels the combination of time and place as fortuitous because those events had good outcomes. Our tracing of the rise of transformative innovations shows that they occurred over a long time, sometimes hundreds of years, during which they built up slowly. Critically, an environment of openness, empowerment of individuals and intense competition fostered the rise and adoption of such transformative innovations. Visionary leaders sometimes charted their course. At other times, innovators endured great hardships and risks to achieve the end result.[22] There was not one swift fortuitous moment and place where innovative success suddenly occurred. Innovation is neither random nor serendipitous. Instead, it is the fruit of conducive conditions in the environment,[23] characterized by these three institutional drivers: openness, empowerment and competition.

Another plausible rival theory to ours for the rise of dominance of nations is that of colonization. The reason for this is that the rise of nations coincides well with the expansion of these nations, frequently through the acquisition and exploitation of colonies. Rome, Venice, Portugal and England became great nations around the time they became great colonial powers. China was not necessarily a colonial power in the strict sense of the word.

However, during the period of China's dominance, it did control vast provinces as part of the empire, if not as colonies. The position of the United States is controversial. Some might claim that the rise of the United States coincided very closely with its westward expansion into Louisiana and other Midwest states, and the annexation of Texas and other Western states. Moreover, the United States did acquire some colonies, such as the Philippines. Thus, territorial expansion rather than the adoption of innovations could be proposed as a cause of its rise.

Territorial expansion and colonization are likely to help in the rise of nations. This factor can be seen most dramatically in the enormous wealth in terms of gold and silver that the Spanish empire extracted from the American colonies between 1600 and 1800. It can be seen in England's crushing the manual spinning and weaving industry in India and forcing the whole country to be a captive market to England's budding mechanized textile industry. However, our argument is not that geographical colonization has no role in nations' rise to power, but that innovation came first. A nation needs some measure of superiority to make another region a colony. That superiority comes from the development and adoption of innovation, likely a transformative one. For example, for Venice it was the galley mass-produced in the Arsenal. For Portugal it was the caravel. For the Netherlands it was the *fluyt*. England and not India mechanized cotton and weaving, and developed the steam engine and steam navigation. Additional details in subsequent chapters will show that innovations came first, followed by national advantage and conquest. Only then did colonization take place.[24]

The limits of the colonization theory can be noted in the cases of the United States, England and Spain. The United States crossed England as a major world power around the end of the nineteenth century, when England still clung to most of its colonial empire. As argued above and detailed in Chapter 11, this crossing was driven primarily by the United States adoption of mass production in numerous industries from agriculture to manufacturing to consumption. During this time period, the United States had far fewer colonies than England. Spain's relative decline started during the eighteenth century, while it still held its colonies. Further, the decline may have been triggered by exploitation of the colonies at the cost of internal innovation and investment. Spain relied heavily on the extraction and import of gold and silver from its colonies. It used the proceeds to wage wars in Europe, with very limited investment in innovation and industrialization on the mainland.[25] The result was hyperinflation as excess gold and silver chased limited goods. Thus, colonial exploitation at the cost of innovation may have been the cause of Spain's decline.

What about long technological cycles? Beginning with the work of Nikolai Kondratiev and Joseph Schumpeter, authors such as Christopher Freeman, Richard Foster and others have proposed that technological change occurs in long cycles that may be S-shaped.[26] After a period of gestation, a new technology bursts forth on the economic frontier accompanied by huge investments, large numbers of entrants, great economic growth and huge benefits to consumers. However, after a period of rapid growth, technological progress reaches certain limits, resulting in stagnation if not recession until the next technology bursts forth. The above pattern has been applied primarily to modern industrial economies since the nineteenth century. Consider for example the economic cycles that coincided with the introduction of the steam engine, electricity, automobiles, the computer or the Internet. While interest in such long technological cycles has a rich literature, recent research in economics has moved away from focusing on such cycles and toward institutions.

Regardless, our thesis does not in any way contradict technological cycles. If at all, it may run complementary to such cycles. However, our thesis differs from that of technological cycles in three ways. First, our interest spans the last 2,000 years, whereas that of technological cycles spans about the last 200 years. Second, the theory of technological cycles suggests certain deterministic patterns in the periodicity of cycles. In contrast, we identify the rise of a new transformative innovation as the cause of the decline of the expansion from a prior transformative innovation. Third, while the theory of technological cycles focuses on the dynamics of technological change primarily within a nation, our focus is on how transformative innovations led to the rise of one nation and the decline of others. Relatedly, while the theory of technological changes focuses on investments and regulation within a nation as drivers of technological cycles, our focus is on institutional drivers that cause a new transformative innovation to emerge in one nation and not another, even though the latter gave rise to the prior transformative innovation. The three institutional drivers responsible for the emergence of transformative innovation in a nation are openness, empowerment and competition, which substantially vary across nations.

Dramatically different from the focus on technological cycles, the World System paradigm focuses on social aspects and relationships. Associated primarily with Immanuel Wallerstein,[27] the World System is a well-established scholarly social science tradition that explores the relations between weaker and stronger countries on economic, social and cultural dimensions.[28] The World System paradigm suggests a global, country-based hierarchy of wealth and power. It assigns countries to core, periphery and semiperiphery countries within a global network. Core countries possess high social status and a large, developed and complex economy. Periphery countries possess weaker social status and typically have weaker economies. They lack the economic capabilities of core countries and are more limited in their ability to fund infrastructure development, education and welfare. As such, the powerful core countries take advantage of the weaker poor countries in the periphery of the System. A basic tenet of the World System paradigm is that a country's core/noncore position remains stable over long periods of time.[29]

In contrast to the World System paradigm, this book shows that positions of power are anything but stable: transformative innovation repeatedly propels new nations into positions of power and brings down dominant nations that are not innovative.

The prior discussion shows how our explanation differs from other explanations for the rise in the dominance of nations, such as geography, natural resources, climate, Christianity, Western culture, colonization, luck or technological cycles. Here we merely sketch out the differences in explanations. Subsequent chapters detail our thesis and provide substantial evidence in support.

Lessons

The research presented in Chapters 2 to 12, of the last 2,000 years of the history of innovation, suggests eight generalizations or patterns that carry lessons for policy makers and industry and world leaders. Here we merely list these patterns. Chapter 12 elaborates on them. However, the reader may fully appreciate these lessons only after perusing all the chapters of this book.

1. *Irresistibility of Innovation.* No single nation can stop the path of innovation.
2. *Transformation from Innovation.* Certain innovations change the economics of cities, regions, nations and the world, transforming small cities, regions, or states into world powers.
3. *Transience of National Domination.* No nation's dominance in the world is permanent.
4. *Disruption from Obstruction.* A dominant nation that fails to extend the innovation frontier falls behind despite its wealth and power.
5. *Curse of Success.* Success sows the seeds of failure.
6. *Transience of Resources.* Geography and natural resources provide a passing but not perennial advantage to innovation, dependent on technological innovation.
7. *Limits of Religion, Geography and Luck as Causes.* Prior explanations for the rise to dominance of nations, such as religion, geography or luck, crumble when one considers 2,000 years of dominance of various nations.
8. *Preeminence of Institutional Drivers.* Three institutional drivers were prevalent to varying degrees when transformative innovations took off, and absent when they did not: openness, empowerment and competition.

2

Roman Concrete: Foundations of an Empire

Around two thousand years ago, several city states and nations flourished around the Mediterranean. However, only one of them rose to become an empire that stretched from England to the Middle East and lasted hundreds of years. What factors led to this greatness? This chapter argues that concrete served as the foundation of the Roman Empire. It enabled the building of harbors, aqueducts, walls, roads and numerous other structures for living, trading and worshiping. These structures enabled the expansion, administration and the flourishing of a vast empire. Concrete played a greater role in the Roman empire than in any other prior empire. Factors that encouraged the innovation of concrete and the stability of the empire were openness to diverse peoples, empowerment of these peoples through a common law and citizenship and competition to outdo standards set by rival nations.

Despite its tremendous contribution to the rise of this empire, historians have paid inadequate attention to this groundbreaking Roman innovation of concrete. Once this powerful innovation declined, the great empire itself faced considerable challenges and began to decline as well. This chapter first details the innovation and then the factors that led to its development and adoption.

The Transformative Innovation of Roman Concrete

What Is Roman Concrete?

The three components of Roman concrete were aggregate, mortar and facing. The aggregate was composed of stones, terra cotta, or tiling and represented the bulk of concrete constructions. In fact, the Latin name for Roman concrete is opus caementicium, after *caementa*—rough, unhewn quarry stones that reminded the Romans of the fist-sized pieces of rock that were used for aggregate.[1] Aggregate could be composed of natural materials, but builders' waste was the most common source. Mortar was created by mixing lime, water and a specific type of sand or ash. The main ash utilized in the production of concrete was *pozzolana*, a volcanic earth that reacted with the water and lime to create an extremely strong and waterproof chemical bond with the aggregate.[2] The chemical bond was a calcium aluminum silicate that is the primary compound of concrete even today. The calcium came from lime, while the silicon and aluminum came from pozzolana. In other parts of the Roman Empire where the volcanic ash was not as readily available, pozzolana was substituted by a similar substance—crushed and burned tile and brick. The brick was made of clay, hardened in a kiln to create a silica compound like that in pozzolana.[3] The term "pozzolan" was first used to describe the ash found near Naples. But a similar volcanic ash—and other pozzolanic materials—were used in other parts of the world, including Greece, India and Egypt.[4]

However, in contrast to these other civilizations, Romans used concrete innovatively by pouring or tamping it into a mold formed from wood. Prior civilizations used concrete (if at all) to join one stone block to another or as a protective coating over brick or stone. Typically, concrete was poured or tamped into a wooden form and then covered with facing.[5] Facing was made of rubble, tiles or bricks and was used to further contain the core while it settled and hardened. Concrete construction consisted of precise mixtures of aggregate and mortar in a form defined by facing or earth.[6] Several factors contributed to the efficiency of Roman concrete.

First, the skilled labor and tools necessary for pouring concrete into molds was substantially less than that involved in laying bricks or stone block upon block. Unlike walls built entirely of bricks or stone, only a single, narrow line of brick facing was necessary for building with concrete. Because the strength of the wall came from the concrete core and not the brick facing, the skill required to carefully and properly position the bricks was not as important as it was in purely brick walls.

Second, the time required for brick facing was much less than the time required for building a thick wall made solely of brick or stone. The innovation was not only in the proper ratios of ingredients for the concrete but also in the simplification of the process of building with concrete. Workers would place alternating layers of mortar and aggregate inside the facing, and the layers would sink into each other to form a solid core. Shovels, rakes and labor were needed, but very little skill was required. This style of building was very different from old-style masonry construction, which required skilled craftsmen to fashion brick or stones, and then heave them into place to construct walls or pillars.[7] With concrete, workmen did not have to dedicate long stretches of time to the task as they did with stone construction.

Third, poured concrete was structurally at least as strong as stone while also cheaper, faster and more malleable for building than stone masonry.

Fourth, concrete walls were "monolithic," like a block of stone, rather than brick-like, so they could be built thinner than masonry walls. Due to these advantages, Roman poured concrete obsoleted the millennium-old method of building stone block upon stone block.

Poured or tamped concrete was good for a variety of architectural forms like residences (palaces and forts), public buildings (temples, markets and arenas) and infrastructure (harbors, roads, aqueducts and walls) (see Figure 2.1). Eventually its use spread to all sorts of construction, from public bathhouses to private homes.[8] This benefit was especially relevant in large cities like Rome and Ostia where space was at a premium during the imperial period.[9] Concrete's structural and economic properties drove a ferment of architectural styles, types and forms all over the Roman Empire, allowing for innovative new techniques to suit local needs.[10] Parallel scientific advances in math and geometry enabled Roman architects to find creative new uses of concrete.

Figure 2.1   View from below of dome of the Pantheon

Source: Rob van Esch/Shutterstock.com.

The most important one of these was vaulted roofing, which was firmly established in central Italy by the mid-first century BCE. Vaulted roofing meant that the ceilings were rounded, like a series of connected arches.[11] When used in domes and vaults, concrete ceilings were lighter than those made of stone and brick and, subsequently, did not need many walls and columns to support them. This allowed for the construction of much larger and sturdier covered spaces, such as the Pantheon.[12] All these creative uses of concrete diffused to all parts of the empire.

The Evolution of Roman Concrete

The innovation of Roman concrete was a long process, in which a distinct form slowly evolved out of older building techniques like the stone and mortar, or plain stone construction that had been used throughout the Mediterranean world for centuries. In fact, there are several prominent examples of pre-Roman concrete, although it was mostly used for flooring. The Minoan Palace of Phaistos, constructed around 1700 BCE in Crete, had concrete flooring, as did the royal complex of Alexander the Great. Roman concrete was different in terms of the material used in the compound as well as the way in which it was used. Instead of simply poured concrete on the ground for flooring like its predecessors, Romans used concrete to create the form in buildings and for architectural support.[13]

Still, its development in Rome was evolutionary, not instantaneous. One of the Romans' biggest innovations was their mastery of the formulas for concrete. Through trial and error over decades and centuries, Romans derived extremely strong and efficient variants of concrete that were good for various ends.[14] By the mid-first century BCE, the Roman architect Vitruvius was able to give precise instructions for how to make the best kind of concrete that was known at that time in his classic *De Architectura*, reflecting almost 300 years of experimentation.[15] He provided specific ratios of lime and pozzolana for the mortar used in the construction of buildings as well as ratios for underwater construction. These ratios are extremely similar to the ratios used for underwater construction today.

Concrete was usually classified by its facing, which evolved chronologically in three clear stages. The original manipulators of Roman concrete confined their core with *opus incertum* facing, composed of small rubble blocks mortared together to make preliminary walls. These could be built up very high and then filled with core. At the end of the second century BCE, Roman builders switched to *opus reticulatum*, which consisted of relatively small square tiles set diagonally. Opus reticulatum could not stand up on its own, so beginning with this transition the core and facing had to rise simultaneously on Roman construction sites. Still, despite the construction complexity, this new type of facing had the advantage of speeding up the building process. This technique, while an improvement, did not last long. From the mid-first century CE, *opus testaceum* took over and dominated concrete construction. Opus testaceum, or burned brick facing, consisted of very flat Roman bricks that had already proved their strength after being exposed to the elements (e.g., roofing tiles) and were then mortared together to form the facing. This type of facing was the lightest, easiest to handle and most flexible yet. It could be combined with other materials to frame openings for windows. It could also be used for both roofs and walls.[16] Opus testaceum is still ubiquitous in surviving buildings. In fact, visitors to Rome today are often surprised to see that ancient Rome appears to be built entirely of brick. Another less-known but equally significant concrete used in the empire is opus signinum—a hydraulic concrete used primarily as a waterproof coating that lined the miles of aqueducts and the 1,352 cisterns used for storing water in Rome.[17]

Improvements in facing were far from the only advancements that allowed concrete to become a highly systematized industry by the first century BCE. There was an active process of architectural innovation occurring in Rome from the third century BCE through the first century CE. It was possible by advances in the technology of concrete. Innovations such as the corbel, metal tie bar and lintel arch, were all architectural features invented by the Romans and used for better manipulation of concrete. These architectural features allowed concrete to be combined with other materials, such as stone and metal, to take advantage of their strengths.[18] In a sense, this was a primitive form of reinforced concrete. The Romans promoted variation and development of their buildings and styles. However, they prioritized efficiency and a positive relationship with the environment. Specifically, they experimented and derived new techniques and ideals to promote the efficient use of water in the process of making mortar and a lime delivery infrastructure. Improving their lime delivery infrastructure facilitated a higher quality mortar in the first century CE than had been used 200 years earlier.[19]

The most important improvement, however, was in the understanding of pozzolana and its qualities. Pozzolana was the major reason that Roman concrete was such a good building material. First quarried near Pozzuoli in the Campania region of Italy, its chemical properties made mortar very strong—particularly because of the interaction between the silica aluminum in the pozzolana and the calcium in the lime.[20] Further, pozzolana mortar could be set underwater.[21] This characteristic enabled, in turn, the extensive use of concrete in the construction of harbors. The Romans understood pozzolana only to a degree—because they did not fully understand the nature of volcanic ash and why it was so different from ordinary dirt. They did speculate that heat underground caused the earth on top of it to be useful for mortar. Vitruvius wrote in *De Architectura* that pozzolana is a kind of powder, which produces astonishing results from natural causes; he speculated that it "must" be because of the hot springs around the area.[22] The Romans' initial lack of understanding of how pozzolana worked is reflected in Pliny's *Naturalis Historia*, where the esteemed naturalist expressed surprise that "dust" (the term he used for pozzolana) could become an effective barrier against the waves.[23] The quality of the pozzolana that the Romans used improved over time as a result of trial and error. At first they used pozzolanella, which can be easily quarried in the open air. By the first century BCE, they were using pozzolana nera and the more high-quality pozzolana rosa, both of which had to be quarried underground. They also got better at removing the inert soil that was mixed in with the pozzolana underground.[24] As a result, the mortar the Romans used in the imperial era was substantially stronger than that which they used in the original Roman concrete constructions.

Open trade provided the materials needed for the development of concrete. As trade evolved, so did concrete. The clay needed for brick facing was abundant in the Tiber and Aniene river valleys. Tall fir trees from the Apennines and southern Italy provided timbers that were needed for scaffolding on large projects and fuel for lime and brick kilns. The lime to put in those kilns was extracted from sedimentary stones like limestone and travertine that were abundant in central Italy and the Apennines. Most importantly, volcanic districts to the north and south of Rome were a ready source of pozzolana.[25] Rome's excellent road system, a fruit of advances in concrete technology, allowed for open transportation. The roads, along with the Tiber and its tributaries, provided an easy and relatively cheap way to transport construction materials where needed.[26]

The introduction and use of improved formulas for and methods of using concrete reflect the Roman spirit of openness, experimentation and learning. Openness to trade allowed Romans to reap the benefits of consolidating appropriate materials from various parts of the empire.[27]

Fruits of Innovation

The innovation of concrete resulted in an explosion in construction and provided social and economic benefits for the people and political benefits for the rulers and the empire.

Building Explosion

By the second century BCE, the development and intensive use of Roman concrete created a boom in construction. The wealth entering Italy, starting in the second century BCE from foreign conquests, provided an increased budget for publicly financed buildings; rulers wanted to use this wealth to build monuments honoring their greatness. The military conquests of the Roman emperors, as well as their desire to increase their public presence, caused a major rise in the funds available for building.[28]

Concrete buildings permeated Roman society. While concrete was far from being the only building material used throughout the Roman Empire, it gradually became a popular construction element. In most cases, it was combined with other building materials to generate strength and aesthetics. Appendix 2.1 illustrates in chronological order some important structures built with different media.[29] It shows the increasing role of concrete over time.

Several of Rome's buildings, all taking advantage of concrete, impressed the world. The Sanctuary of Jupiter Anxur, built in the second century BCE just south of Rome, bears piers and arches supported by squared pillars. The entire structure is made of concrete with stone facing. This building marked the beginning of combining concrete with structural arches and vaults.[30] Next, the Sanctuary to Fortuna at Praeneste set a precedent in Roman construction, as it amalgamated concrete and stone columns.[31] Setting the record as the largest concrete building as of the second century BCE, the Sanctuary of Primigenia displayed aspects such as concrete and stucco veneers, limestone and innovative seven-vaulted terraces. It is widely considered an expression of the Roman Empire's success in architecture.[32] Finally, the best-known Roman building, the Colosseum, is also built of concrete. Constructed at the end of the first century, the Colosseum has numerous arches and vaults, all of concrete.[33] It is symbolic of Roman architectural prowess. Still, its architectural style also expresses openness to and dialogue with Greek architecture. It uses Doric, Ionic and Corinthian columns that juxtapose the arch and vault design at the bottom, middle and top levels respectively. The Doric columns at the bottom level represent a strong Greek influence. The Corinthian columns at the top level, while originally Greek-Hellenistic in style, later became synonymous with Roman architecture. This bottom-up architectural approach is a good example of openness-competition in architectural design: it accepts Greek influence but also places the Roman notion at the top, symbolically a superior position.

Rome's greatest architectural accomplishment, the Pantheon, remains a turning point in the use of concrete. Its numerous levels are each built with concrete aggregate mixes that differ in weight in a lightness-ascending order. It addition, it is unique in incorporating niches and cavities to reduce weight. The result is the astounding height of what is, still today, the world's largest unreinforced dome.[34] Its Hellenistic and Greek design influence represents Rome's openness to foreign styles on the one hand and the superiority of innovative Roman architecture on the other hand, in a continuous architectural dialogue of acceptance-competition.

Though concrete seems to have been first developed for "buildings," it was widely used for engineering works as well. It was a key material in the construction of roads from the Imperial Forum to the long roads that connected the city of Rome to other parts of the empire.[35] It was also used by the military to quickly construct fortifications around cities and forts under siege. After the first century BCE, concrete was the main ingredient for aqueduct construction—a major element in the empire's water supply system.[36] The piers surrounding the aqueduct would often be faced with mortar and brickwork that encased the concrete interior. In addition, concrete was used for harbor construction. The ability of pozzolana mortar to set underwater allowed the Romans to use concrete construction to build deep-water harbors. They would make enormous prefabricated concrete units nearby, load them onto boats and drop them directly into the ocean in the desired location. Emperor Claudius, for example, shipped pozzolan 162 miles from Pozzuoli to build a harbor at Ostia[37] and sank prefabricated concrete obelisks to assist in the construction of the port there.[38] These harbors allowed Roman merchants to use boats of a size that could not enter a river mouth or shallow coast.[39] The most famous example of this is Herod's harbor at Caesarea, in the province of Judaea, where he also built many temples and entertainment facilities out of concrete.[40]

The Roman transition from republic to empire was completed with the rise of Augustus in the first century BCE. It was accompanied by an enormous surge in construction. Buildings and infrastructure were erected at a furious pace for the next approximately 200 years, as emperors, generals and noblemen attempted to leave their mark on history, to please the public and to accommodate the empire's rapidly growing population. From the city of Rome to the most distant provinces of the empire, monuments such as the Pantheon and Herod's port city of Caesarea in Judaea still stand today as evidence of this unprecedented building boom. This boom was made possible by the innovation of Roman concrete. The Romans considered this material transformative because it allowed them to build stronger and more elaborate structures at a faster pace, using less skilled labor and at a lower cost than any other method of construction that had previously been employed. Specifically, concrete enabled stronger walls, roads, bridges, domes, buildings, aqueducts and harbors. Developed and honed for some time, Roman concrete construction was well established as a technique just in time for the Augustan period, at which point its use exploded in Rome and its provinces.

Social and Economic Benefits

The engineering projects facilitated by concrete had significant economic and social benefits. Public roads covered the entire empire with the exception of some outlying regions such as Britain north of the wall.[41] Better roads allowed for easier and cheaper transportation around the empire, an invaluable asset for Roman rulers. They used them to unite and consolidate military conquests and subsequently to aid in controlling the vast terrain under their rule. This system also aided trade within the empire, especially the transportation of goods from border areas to the center of the empire. It also facilitated the diffusion of new styles and techniques from the center of the empire to border areas. The importance of a good road system for imperial maintenance has been recognized ever since.[42] The construction of deep harbors, along with the trade routes, facilitated trade and hence the diffusion of ideas and innovations as well as architectural and engineering innovations. The ability to harbor bigger ships allowed Roman merchants to take advantage of economies of scale and become more efficient, thereby improving trade and boosting the empire's economy.

The growth of Roman society during this time, along with the great fire of 64 CE in Rome, required more housing. It also encouraged Roman authorities to construct more public buildings like bathhouses and entertainment centers. Concrete facilitated the expansion of these public services, while the method of concrete vaulted roofing allowed for large covered gathering places, like bath buildings, markets or fora. In addition, concrete roofing allowed for efficient ways to accommodate large audiences, such as in the Colosseum.[43]

The economic benefits of some large projects like baths, aqueducts, amphitheaters and roads were spread to various classes of people in the city. For example, in an analysis of the construction of the Baths of Caracalla, researchers suggested that almost 80 percent of the immense cost of the Baths went to the workforce, which would have represented a significant economic boon to people in Rome as well as the territories that were the source of raw materials.[44]

Political Benefits

Concrete gave the Romans the ability to build at a high volume and a low cost, greatly benefiting the growth and maintenance of state power and stability. Through public construction, Roman rulers were able to spread patriotic values crucial to imperial continuation. Indeed, scholars consider Roman buildings to bear underlying meanings behind their construction. Their buildings emanate an expression of unity, endurance, authority and pride. As such, monumental architecture was a manifestation of Roman values and thus facilitated submission to the empire, obedience to the emperor and imperial cohesion.[45] Monuments and temples therefore served the continuity of the Roman Empire. Citizens were inspired to feel proud of their collective identity, whether city, province or empire, by the imagery and scale associated with public buildings. Research has shown that tall buildings symbolize economic growth and opportunity in people's minds.[46]

Upon conquering a new territory, Romans constructed roads and bridges that led to the capital, as well as aqueducts that transported water into cities' centers. The monumental aqueducts and the fresh water that they brought, were reminders of the power and generosity of the Roman government.[47] The physical aspect of the building was as symbolic and meaningful to the people in terms of strength and opportunities associated with the sovereign as was the content of the building, which conveyed aspects of Roman culture. Further, the fact that these public buildings so often existed for the benefit of the population— such as bathhouses and entertainment centers—would have increased the public's satisfaction with their government. That, in turn, served the government as well.[48] Public entertainment in ancient Rome was often a form of political communication; leaders would hold games such as gladiator matches to curry favor with the public and send messages to it. These games, taking place in breathtaking concrete buildings, were simultaneously a demonstration of—and an active pursuit of—power.

The development and use of concrete facilitated urbanization in Rome and other cities around Italy. Vitruvius noted in *De Architectura* that because of concrete construction, the Romans were able to build taller, more economical buildings at a fast pace to accommodate more people in the city.[49]

Drivers of Innovation

Three institutional drivers played a key role in the development of and use of Roman concrete: competition, openness and empowerment.

External and Internal Competition in the Roman Empire

Competition that drove innovation across the empire was internal and external. The external competition was with other nations. The primary one was the Greeks, the competent predecessor rulers of large parts of the pre-Roman world. Another type of competition that drove some consequent evolution in architecture was internal and involved competition among rulers. In both cases, competition was with predecessors as well as with contemporary opponents.

External Competition

The Romans ceaselessly competed with the Greeks in regard to art, architecture, technology, culture and empire. Much of this competition began because of the contact the Greeks, Etruscans and Romans had through trade and conquest at the turn of the first millennium. These conquests resulted in Rome taking over many areas formerly considered Greek. The Greek-Hellenistic culture left a very compelling heritage in arts, architecture and culture. Thus, with early trade patterns and as the conquests of small cities in Southeastern Europe and North Africa (like Carthage) increased, the Romans adopted more Greek styles, tactics and technologies.[50] This adoption was a sign of Rome's openness, but also expressed Rome's sense of outdoing a perceived superior predecessor. Greek accomplishments inspired Rome's emperors and government to rival the Greek culture and surpass it by adapting and improving on Greek concepts and accomplishments. The Romans made efforts to exactly copy works of Greek art; these replications only begin to explain the Greek influence and the underlying competition, which drove some of the Roman innovations.[51]

The attitude toward Greek architecture stood in contrast to Rome's attitude toward Greek art. While Greek art was merely replicated, Greek architecture provided a basic platform for the massive evolution in architecture that Roman builders developed. While copying Greek art represented Rome's openness, advancing Greek architecture represented Rome's sense of competition and its strong desire to surpass its competent predecessors.

Improvements in concrete construction, an enlarged population and increased wealth fueled Roman competition with the accomplishments of other nations. The Romans used concrete to show their ability to conquer and reshape their natural environment, demonstrating Romans' power over the material world. The then "Seven Wonders of the World" had a similar influence in Rome, as leaders strove to match what were known as the greatest buildings in the world. Nero even attempted to recreate some of them in Rome himself. Some scholars suggest that he erected a 120-foot statue of himself to imitate the Colossus of Rhodes.[52] He also built gardens elevated on concrete vaults at the Domus Tiberiana, in what some scholars have argued was a replication of the hanging gardens of Babylon.[53] His love for lavishness, luxury and egoism was prevalent in his rebuilding of Rome after the fire. Scholars are undecided whether Nero caused the fire himself or whether it was due to other causes. In either case, the palace Nero built after the fire towered over Rome in lavishness and grandiosity. His competitive quest for greatness drove this lavishness and grandiosity in his palace and other buildings he constructed.[54]

Internal Competition

Roman emperors tried to compete with their predecessors and successors in building extravagant and memorable buildings. Some publicly financed building projects became a sign of imperial generosity and allowed rulers to promote their names and increase their stature.[55] Projects funded by the emperor himself particularly exhibited this phenomenon. Starting mostly with Julius Caesar, the Romans began to deify the emperor and create a cult surrounding him, which helped maintain imperial power.

Augustus took this one step further, creating a full-fledged cult around himself, thus positioning himself as superior to former and possibly future rulers. His construction of public works and monuments served as propaganda, bolstering his popularity and the aura of the Roman Empire. The Roman government and its leaders used buildings to maintain popularity, assert dominance, awe subjects and establish a sense of unity in the empire. Public monuments and buildings were also erected to celebrate momentous imperial events such as ascendance to power and victory over enemies.

For example, the death of both of Augustus's heirs had left some doubt as to who the next emperor would be. Eventually Tiberius emerged as the clear right-hand man to the emperor, but rumors persisted about splits between him and Augustus. To quiet any public doubt, Tiberius restored an ancient temple and rededicated it to the unity of Augustus and himself. Then in 12 or 13 CE, he dedicated an important monument, the altar of the Numen Augusti (Genius of Augustus). As the first great monument for the worship of the emperor, it served to legitimize Tiberius's power in light of the fierce competition with his opponent. Indeed, following Augustus's death in 14 CE, Tiberius succeeded him. These events exemplify how competition among rulers and potential successors contributed to the further development of construction.

Fires frequently plagued Rome and thus presented opportunities for emperors to rebuild the city as they saw fit—facilitated by the fast, economical and somewhat more fireproof method of concrete construction. For example, in 64 CE, a fire struck Rome and destroyed much of the city. That fire became a symbol of Nero's reign and the initiation of Nero's attempts at rebuilding Rome.

Following the lead of the imperial family, several other nobles and patrons also sponsored the construction of public works to increase their popularity. The Roman people as well as aristocrats understood such demonstrations of power.[56] For example, Herod, the ruler of the autonomous province of Judaea, used construction for self-aggrandizement. He built Caesarea as the showcase of Judaea and sponsored buildings all over his province and the empire in general, including gymnasia, theaters, temples, marketplaces and refurbishments of fortifications and plazas.[57] Herod's power stresses the important role of construction in internal competition among rulers in the building of the empire.

Roman Openness to Peoples and Ideas

The major expression of openness was the Romans' tendency and willingness to adopt, absorb and advance the achievements of the peoples and nations it conquered. An important manifestation of openness was trade.

Trade played a major role in the empire. The Romans realized the economic importance of trade, and rulers supported trade within and outside the empire. Trade kept the idea market vivid and intriguing. The Romans, Etruscans, Egyptians, Spanish and Greeks all traded with their counterparts in Europe and Asia. Trade occurred due to a constant demand for necessities and luxuries, and a desire to travel and explore.[58] Rome's trade routes with other societies increased over time with imperial strength. In Rome's early days, trade focused on military goods necessary to consummate the society's infrastructure and government. However, as the empire grew and boundaries stretched, trade shifted toward relations with Scandinavia, China and along the Silk Road.[59] The type of traded goods also changed. For example, China's Han dynasty traded desired luxuries with the Roman Empire.[60]

The two most common imports from China were silk and spices. Spices traded mainly from China and other regions in southeastern and southern Asia included clove, cardamom, cinnamon, pepper, cassia, ginger and nutmeg. Spices made such an impact on the Romans that a portion of their city became known as the "Spice Quarter."[61] Silk made a similar impact. A rumor even began that one pound of Chinese silk brought on the Silk Road was worth as much as one pound of local gold.[62]

This exchange in luxuries, which increased over time, became a natural platform for the exchange of ideas and concepts. With trade and travel came ideas about science, art and architecture. It also facilitated openness to absorbing and adapting the achievements of others, while also promoting the diffusion of innovations across the empire. Thus, openness led to the adoption and diffusion of new ideas. When it came to architecture, styles and techniques diffused along imperial trade routes. Openness affected the innovation of Roman concrete. Trade enabled the Romans to use others' ideas as platforms for their own creativity and allowed the import and export of essential materials for the manufacture of concrete. Once foreign ideas were adopted by the Romans, they became grounds for further evolution and advancement. Architectural fashion changed rapidly and novel styles emerged. Some coexisted with old ones, while others rendered old styles obsolete. Rome revealed a confident attitude of taking other nations' concepts and applying a "pragmatic" and exploitative approach to making them their own.[63]

Empowerment through Citizenship, Mobility and Incentives

The major expressions of empowerment were citizenship, social mobility through meritocracy and incentives. The rulers offered citizenship, with increasingly liberal rights, to various peoples and some slaves in the regions they administered. Over the history of the empire, this citizenship included increasing rights and privileges. With citizenship came the ability to maintain an authority position in the army, to enter various administrative positions and to rise through the ranks based on merit. Such meritocracy enabled the empire to draw upon the talents and energies of an increasingly diverse population. All this allowed for creativity and innovation. This section explains empowerment in terms of citizenship, social mobility and incentives.

Citizenship

Citizenship was one of the biggest means of empowering individuals in the Roman Empire. Desired by Romans and non-Romans alike, Roman citizenship afforded individuals a slew of important rights—these rights were divided into public rights (*iura publica*) and private rights (*iura privata*). Public rights included the right to vote in popular assemblies (*ius suffragii*), the right to hold public office (*ius honorum*) and the right to appeal to Roman courts (*ius provocationis*).[64] There were two main private rights: marriage right or the right to contract a lawful marriage (*ius connubi*) and property rights or the rights to acquire, transfer and hold property (*ius commercii*).[65] The latter of the two was particularly significant, as wealth was measured largely by the amount of property one held. The right to amass and keep wealth motivated individuals to buy, trade and develop land, as well as design, engineer and construct monuments to please the leaders. The right to reap and keep rewards spurred innovation. Roman citizens were also exempt from taxes collected to maintain the Roman army and to commission public work (*tributum*).[66] In exchange for these rights, citizens had civic obligations that varied over time but were largely required to serve in the army and pay a separate set of taxes.

As the Roman Republic grew into an empire, the meaning of citizenship expanded slowly to a greater portion of the empire's population. Initially the other Latin states, as well as other territories on the Italian peninsula, were considered municipia—territories of a lower status in Rome inhabited by secondary citizens (*civitas sine suffragio*).[67] These citizens could enjoy the private rights and rights to appeal to Roman courts but were not allowed to vote and, accordingly, hold office.[68] The first step toward increased enfranchisement was in 88 BCE, following the Social War. The Social War was waged between the Roman republic and several other Italian cities that had previously been Rome's allies. The Italian cities that protested were previously denied full citizenship and pushed for independence from Rome.[69]

The war resulted in a Roman victory, and Julius Caesar proposed Lex Julia, a law that extended citizenship to all Latin and Italian communities that did not revolt. It was followed by Lex Plautia Papiria, which that granted the same opportunity to citizens who had been excluded in Lex Julia.[70] The Romans granted their Latin and Italian allies full citizenship. Lex Julia also set the precedent for granting citizenship to those who had distinguished military careers.[71] The enfranchisement of Latin and allied states was accomplished by 80 BCE.[72]

There were two main ways that individuals could acquire citizenship. Apart from inheriting citizenship at birth, individuals could receive it as a gift from high-level Roman officials. In exchange for loyalty, or by willingly surrendering to Roman legions, local rulers and inhabitants of conquered territories were gifted with Roman citizenship. The territories were developed into municipia, and the local rulers were frequently able to maintain a certain level of authority even after the land was incorporated into the empire and a Roman magistrate was assigned to the region. While the rulers sometimes received full citizenship, most conquered peoples received the secondary form of citizenship. After a certain period of time passed, if those holding civitas sine suffragio proved their loyalty to the empire, they received full citizenship. This process of full enfranchisement was expedited in provinces with high levels of Italian immigrants.[73] Upon earning this citizenship, they could exercise greater political influence and join Rome's public assemblies. The second way to acquire citizenship was in accordance with Lex Julia—non-citizens who served in the army frequently received citizenship for commendable service. This practice of awarding citizenship continued until 212, when the Roman emperor, Caracalla, declared all men in the Roman Empire were citizens.[74]

Social Mobility

Especially with openness to diverse peoples in the empire, citizenship provided opportunities for social mobility regardless of one's background. Citizenship enabled people to rise through administration, the army and electoral offices through a process known as cursus honorum. Cursus honorum was a sequential order of public offices, both military and political.[75] It was designed for men of senatorial rank but was also open to citizens favored by the emperor or those voted into power by the public magistracies.

There are several historical examples of individuals of both Roman and non-Roman origin rising through the ranks. Individuals born outside of Rome because of their citizenship were able to climb political ranks, even so far as to become emperors. After the Julio-Claudian line, of the 36 emperors whose birthplaces are known, 17 were outsiders, living both outside Rome and outside of the Italian peninsula (see Table 2.1). The rise of non-Roman emperors is an indication of empowerment.

A prominent example of this is Septimus Severus, emperor from 193 to 211. Severus was born into a Roman noble family, but in the Roman province of Leptis Magna in Africa.[76] Despite the place of his birth, he was promoted to the level of senator by Emperor Marcus Aurelius and as a citizen, rose through the ranks, eventually claiming power and ruling as emperor. An even more noteworthy example of a non-Roman becoming emperor was Maximinus Thrax.[77] Maximinus was a Thracian who joined the army as a common soldier and rose through the ranks. His rise to power followed military reform by Severus, who allowed promotions for outstanding officers. These reforms, along with Caracalla's edict and his military prowess, resulted in Maximinus

Table 2.1 Region of origin of Roman emperors

Region	Number of Emperors	Total Number of Years of Their Reign
Rome	9	116.9
Italia	12	67.6
Hispania	2	40.6
Libya	1	17.8
Gaul	2	32.8
Syria	2	3.8
Thrace	1	3.3
Caesarea	2	14.2
Serbia	6	32.7
Unknown	9	25.8

Source: *Oxford Dictionary of Byzantium* (New York; Oxford: Oxford University Press, 1991).

being the first soldier to become emperor. Although the senate considered him a barbarian, they gave Maximinus imperial powers.[78]

Besides the army and administration, a means of social mobility in industry was the collegia, or builders' organizations. These were primarily social organizations that were not mandatory for builders to join, but membership was a demonstration of financial and professional success because it required entry fees and dues. Therefore, collegia provided a means of advancement and sense of achievement for the lower social strata involved with construction. For example, joining the brickmakers' collegium was a great way for people formerly of the lower classes to announce their arrival as upwardly mobile citizens.[79] In this manner, Roman construction facilitated a sense of empowerment for individuals from various social strata. There were other forms of social mobility resulting from construction works. Contractors, or redemptors, were from the lower classes and were often freed men or their descendants. Redemptors bought their materials from wealthy landowners, who were usually of the senatorial class. The Roman institution of clientela involved obligations between people of different social strata and created a formal bond between the landowners and redemptors. For example, Emperor Domitian charged L. Paquedius Festus, an imperial contractor, with the task of constructing the Monte Affliano tunnel.[80] These sorts of arrangements enabled both the upper and lower classes to benefit from the building boom.[81] The contracts drafted were explicit and detailed, and addressed accountability and costing.[82] Individuals who engaged with other classes enjoyed social connections and options. The latter empowered individuals, thus driving their further engagement in the evolution of construction technology.

Incentives

Rome employed nonmonetary and monetary incentives to promote the development and use of innovations.

Citizenship was one of the most important nonmonetary incentives that the Romans used to create productive members of the empire and strengthen the ties throughout the vast empire. Individuals were eager to become citizens, in hopes of acquiring property rights and exercising political influence in the magistracies. The large number of citizens and those hoping to earn citizenship ensured that troops were constantly available for military quests and the protection of the empire.

Emperors and senior administrators rewarded local rulers and noblemen for their contribution to the imperial stock of monumental buildings and civil infrastructure. This policy drove local rulers, noblemen and architects to exploit and advance construction technology. For example, after Augustus's reign and between 27 and 96 CE with emperors Tiberius to Domitian, the early empire instigated significant building projects for different social purposes. Theaters, law buildings, basilicas, housing and the like were built to serve specific needs to Roman culture. The builders used the latest and most innovative technology, design and planning.[83]

While, for the most part, the people sponsoring construction received recognition for the completion of the work, sometimes architects enjoyed recognition for their work. One such architect was Apollodorus of Damascus. Apollodorus rose to fame under Emperor Trajan. He was credited with designing most of Trajan's imperial buildings, including the baths, forum and column that bear Trajan's name. He accompanied Trajan on several military quests and advised Emperor Hadrian for a period of time as well. Apollodorus's recognition is an example of Roman openness (as he was a Damascus-born Greek engineer) and the incentivization that architects had to improve their work and innovate. This was a form of empowerment. The scope of construction works and the importance Roman society attributed to them maintained an advanced level of labor based on selection, incentives, rewards and power—all leading to the sustaining of innovation.

Power and wealth were not the only incentives for engaging in construction. Because the building industry in Rome was comprised of people from so many different levels of society, it also served as an excellent network of intraclass ties. Upward social mobility as explained in the prior subsection, served a nonmonetary incentive to perform productively within Roman society. This social mobility motivated workers to utilize their skills and extend their existence beyond their current social class.

Construction served as a big political and economic incentive to the leaders, entrepreneurs and engineers. There were strong incentives to those engaged in the building activity, encouraging them to use new ideas and improve existing technologies. The message was clear: construction was a source of honor, power and wealth. That, in turn, incentivized and empowered individuals to engage in the building activity and promote it.

The Decline of Roman Concrete

Concrete disappeared from use around 300 CE. Other types of masonry replaced it, and nothing like concrete showed up again for about 1,500 years. At that time, the Industrial Revolution in Western Europe brought about the development of modern concrete.[84] Several factors led to the decline of Roman concrete.

First, concrete flourished in the Roman Empire as construction flourished with the expansion of the empire. As new territories, regions and cities came under Roman rule until its peak in 117 CE (see Figure 2.2), the demand rose for massive construction of roads, walls, harbors, aqueducts, sewers, amphitheaters, baths, temples, palaces and homes. However, the empire peaked when

Figure 2.2 Roman Empire at its peak in 117 CE

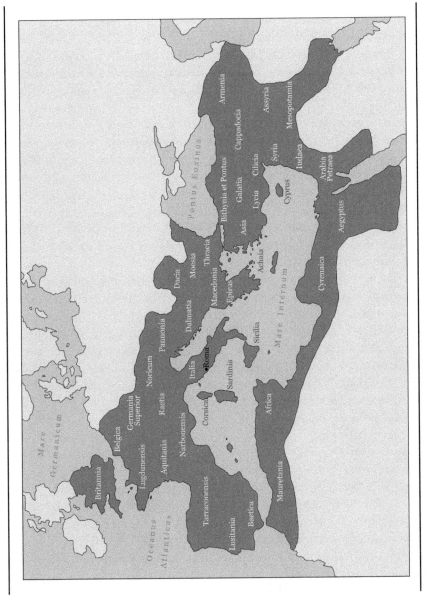

Source: Peter Hermes Furian/Shutterstock.com.

it pushed against some natural boundaries: oceans to the west, rivers in the north, lakes in the northeast, rivers and deserts in the east, and deserts in northern Africa. Once the empire stopped expanding, construction slowed, resulting in less demand and innovation in concrete.

Second, the Roman Empire did not establish a robust norm for imperial succession. As a result, struggle for succession was rampant, with frequent assassinations and civil wars. These struggles diverted limited resources from construction to internal warfare, especially after its peak. Lack of expansion of the empire exacerbated internal conflict, further diverting resources from construction to wars. Concrete construction involved economies of scale in the simultaneous functioning of many large industries such as brick making, lime burning and pozzolana quarrying. The simultaneous smooth functioning of these industries at a high level fueled the boom in construction. The decline in demand for concrete disrupted these industries, leading to a corresponding decline in the use and knowledge of concrete.

Third, Diocletian reforms suppressed entrepreneurship and innovation in construction and concrete because they hindered empowerment. Once Diocletian took over in 284 CE, internal competition and warfare decreased. The reforms of Diocletian, originally meant to improve governance, effectively increased state control over the supply of materials and the supply of labor for building projects. Diocletian overregulated the building industry to the point that it no longer served as a source of social mobility and political or economic empowerment. Building projects no longer had widely dispersed economic benefits.[85] The overgovernance lessened freedom for individuals to build structures as in the past and so lessened competition among local rulers. Noblemen could no longer benefit from construction of public buildings, as little power was associated with such actions. As the government overregulated building, materials and labor, disempowered individuals could not build without permission from high authority and innovation and entrepreneurship in construction declined.

Fourth, around this time, resources were tight. Lack of expansion of the empire meant that there were no new regions to build, tax and exploit. Taxes for construction could be raised on only existing populations. So, these populations viewed large building projects (such as palaces and temples) as burdens they had to fund with taxes.[86] This attitude reduced incentives to build and decreased innovation in engineering and construction. The driving forces of building (prestige and power for the upper classes and social mobility for lower classes) no longer served as incentives for innovation. These changes greatly reduced the level of construction in Rome by 300 CE. At this point, the supply chain that facilitated Roman concrete construction so well during the boom was broken by the lack of demand over the course of the previous century.

Fifth, religious change may have also contributed to the decline of concrete. As Christianity became more prominent, the empire became less open to pagan rituals. Pagan-related institutions and ceremonies began to lose popularity. Gladiator fights, exotic-animal displays and fights, chariot races and athletics lost their appeal. Christian leaders disfavored these practices as "pagan". Amphitheaters lost their prior exotic content to plays with Christian themes. Entertainment venues, such as stadiums, racehorse tracks, theaters and amphitheaters changed their facades and functions. At first, they hosted ritual processions. But eventually, they were simply abandoned. Furthermore, there was no push for the construction of new churches, as old Roman temples were frequently converted to meet the needs of the church. Once the jewel in the Roman construction crown and a major driver for architectural evolution, entertainment buildings simply ceased to be of interest. They were no longer a source of pride, an emblem of imperial cultural unity or of Roman cohesion.

Past its peak, the forces that drove Rome's rise slowly began to ebb. Openness declined. Rome endured heavy threats to its borders by other societies in the mid-third century.[87] No longer as powerful, the interactions with neighboring societies changed from cordial exchanges of goods, ideas or customs of a superior entity to military warfare with a threatened empire. This situation triggered introversion rather than extroversion. The Romans' willingness to adopt and absorb turned to defense and shutting out invasions. The openness that had allowed for a fertile intellectual, cultural and social exchange, decreased, hindering new influences for innovation and technological growth.

Competition also declined. Competition with the Greek culture was forgotten. Competition with concurrent cultures was primarily militant in nature rather than cultural or social. As such, it did not promote architectural innovation. By the time of Diocletian, competition among emperors became destructive. Instead of competing with their predecessors, emperors focused on suppressing rivals. Diocletian introduced the concept of co-emperors and the division of the Roman Empire into the East and West, which further divided resources and increased diseconomies of scale. The eroding economy made it hard to invest in construction. At the same time, the empire did not reward high-profile construction works by local rulers. As the incentives were no longer there, construction had little merit.

Empowerment diminished gradually as individuals involved in construction had less opportunity for social mobility, recognition or power in society. On the way to its peak, the forces of competition, openness and empowerment drove the evolution of concrete. They also drove the consequent evolution of Roman architecture and civil engineering. Once the drivers of innovation weakened, concrete declined with the concurrent reduction of innovations in architecture, construction and engineering.

Once concrete declined, Romans were unable to properly repair public buildings and maintain the infrastructure, or initiate massive construction to disseminate Roman values and control the empire. This situation may have aggravated the decline of the empire. The slow decay of an empire built of concrete is likely when concrete is no longer used. Concrete helped stabilize and maintain the power of the empire at the time when it took off. Concrete's fall accompanied its deterioration. Unlike prior empires, Rome was built on concrete. Once concrete declined, sustaining this empire became hard. Authors have proposed many factors to explain the fall of Rome.[88] The decline of Roman concrete may be another such factor that contributed to the decline of the empire.

Appendix

Table A2.1 Important Roman buildings

Building Name	Construction Period	Main Materials	Emperor of Commission	Style and City
Temple of Jupiter Capitolinus	241 BCE	Marble, stone	Republican Period	Etruscan Temple; Cosa
Sanctuary of Jupiter Anxur	Second century BCE	Tufa, concrete, stone-concrete facing (opus incertum), limestone	Republican Period	Terracina
Sanctuary of Fortuna	Second century BCE	Tufa, concrete, limestone	Republican Period	Praeneste (present-day Palestrina)
Temple of Fortuna Virilis	Second century BCE	Travertine, tufa, stucco	Republican Period	Roman Rectangular Temple; Rome
Temple of Vesta	80 BCE	Concrete, *opus incertum*, travertine, stucco	Republican Period	Roman Circular Temple; Tibur (Tivoli)
Arch of Augustus	Built late second century BCE, Dedicated, 29 BCE	Brick faced with travertine	Julius Caesar (when dedicated)	Etruscan; Rimini
Bridge at Narni	27 BCE	Concrete faced with travertine	Augustus	Narni, Rome
Mausoleum of Augustus	25 BCE	Concrete, tufa, travertine	Augustus	Roman Mausoleum; Rome
Tomb of Caecilia Metella	20 BCE	Concrete, brick, travertine, Pentelic marble	Augustus	Roman Tomb; Via Appia

(continued)

63

Table A2.1 (*cont.*)

Building Name	Construction Period	Main Materials	Emperor of Commission	Style and City
Theater of Marcellus	23–13 BCE	Concrete, marble, stone, travertine	Augustus	Graeco-Roman Theater; Rome
Maison Carrée	16 BCE	White limestone	Augustus	Roman Rectangular Temple; Nîmes
Temple of Mars Ultor (at Forum of Augustus)	14–2 BCE	Concrete, pavement, Carrara marble, multiple stone types	Augustus	Roman Rectangular Temple; Rome
Pont Du Gard Aqueduct	First century CE	Concrete, sandstone	Claudius?	Roman Aqueduct; Nîmes, France
Markets of Trajan	100–112 CE	Concrete, brick	Trajan	Rome
Temple of Venus and Rome	123–135 CE	Concrete, granite travertine, marble	Hadrian	Roman Rectangular Temple; Rome
Hadrian's Villa	124 CE	Concrete	Hadrian	Roman Houses; Tibur (Tivoli)
Pantheon	125–128 CE	Marble, brick, concrete, granite, Pentelic marble	Hadrian	Rome
Temple of Diana	130 CE	Primarily stone	Hadrian	Roman Rectangular Temple; Nîmes
Odeion of Herodes Atticus	161 CE	Concrete, stone, marble, cedar wood	Lucius Verus, Marcus Aurelius	Roman Theater; Athens
Baths of Caracalla	211–217 CE	Concrete, brick, marble veneer	Caracalla	Roman Bath; Rome

(*continued*)

Table A2.1 (*cont.*)

Building Name	Construction Period	Main Materials	Emperor of Commission	Style and City
Aurelian Wall	271–275 CE	Concrete, brick	Aurelian	Porta San Sebastiano, Rome
Triumphal Arch	300 CE	Concrete, brick, marble	Maximian, Diocletian	Roman Arch Thessalonica
Basilica of Maxentius and Constantine	306–313 CE	Concrete, brick	Maxentius/ Constantine I	Roman Basilica; Rome
Basilica at Trier	310 CE	Brick (interior and exterior)	Constantine I	Roman Basilica; Trier
Arch of Constantine	312–315 CE	Marble, brick	Constantine I	Roman Arch; Rome
Temple of Minerva Medica	Early fourth century	Brick-faced concrete	Constantine I	Roman Temple; Rome
Old St. Peter's Basilica	320–330 CE	Marble, timber	Constantine I	Roman Basilica; Vatican City
Interior of Santa Costanza	Mid-fourth century	Brick, granite, marble	Constantine I, Julian the Apostate	Roman Temple; Rome

3

Swift Equine Warfare and the Rise of Mongol Power

Led by Genghis Khan and, later, by his sons and his grandson Kublai Khan, the Mongols crafted an empire between 1206 and 1368 that included much of Asia, stretching from Korea in the east to the Danube in Europe and the Eastern Mediterranean lands in the west (see Figure 3.1). For a time, the empire also included parts of Russia, Poland, Hungary, Syria, Iraq and Persia.[1] It is by far the largest contiguous land empire in the world, covering about 12 million square miles.[2] In 25 years, the Mongols conquered more territory and peoples than the Romans did in 400 years, and five times the area conquered by Alexander the Great.[3] To further appreciate its military prowess, we can compare it to the European and Muslim powers of the time. While the combined powers of Europe struggled to capture a few ports on the Mediterranean from the Muslims in 200 years, within 40 years, the Mongols conquered every Muslim kingdom and city between Central Asia and the Mediterranean Sea.[4]

How did a relatively poor people, numbering around a million, living in a remote part of the world, come to conquer and rule such a great territory, encompassing over a hundred million people? They did so through innovation in rights, communications, administration, trade and especially warfare. These innovations developed through a great openness to ideas, tools and talent from conquered peoples, while empowering them with new rights, including meritocracy, travel and trade privileges.[5]

Figure 3.1   Mongol empire at its greatest extent

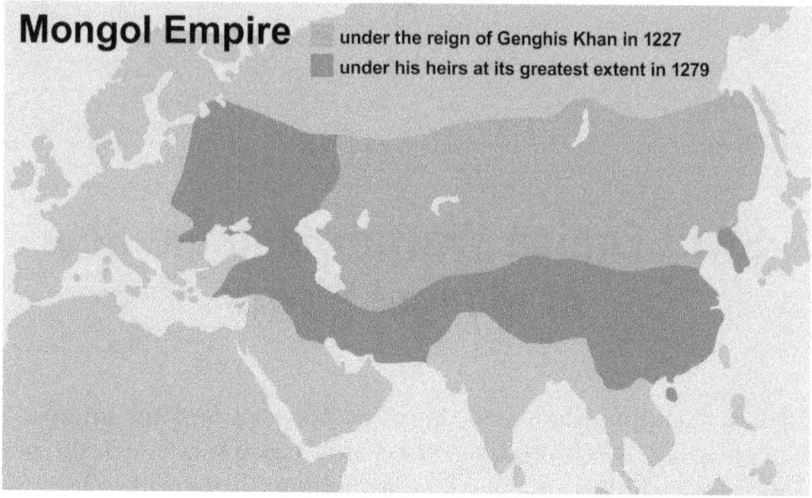

Source: Peter Hermes Furian/Shutterstock.com.

The Mongols were far from being merely a wild, barbaric horde of looters and killers, as some may believe. Rather, the Mongol army was a well-structured, well-disciplined and brilliantly commanded force of warriors using innovative tools, strategies and tactics.

At the root of Genghis Khan's huge military achievements were the traditional Mongolian horses and their riders.[6] From these resources, the Mongols developed a new military technique, which we call swift equine warfare. Both horses and riders had great potential. Genghis Khan undertook the task of transforming thousands of them into a disciplined, structured and efficient fighting force.[7] Led by Genghis Khan, Mongol military leaders added strategic and tactical genius, embraced new tools of war and provided excellent organization and communication.

In addition, Genghis Khan destroyed the old feudal system based on aristocracy and birth. In its stead, he established a new system of individual merit based on a new international law.[8] He lowered taxes for farmers and traders, and abolished them for doctors, teachers, scholars, lawyers and religious leaders. He established a regular census and an exceptionally effective international postal system. While religious dogmatism and persecution was widespread in other parts of the world at the time, he decreed complete religious freedom. Most importantly, he firmly established the Silk Road that strongly linked Asia, the Middle East and Europe, enabling trade and the diffusion of innovations among these three separate regions. Knowledge about innovations such as gunpowder, printing, paper money and the compass may have passed to the West through the Silk Road or other exchanges between East and West facilitated by the Mongols. This exchange may have been one of the stimulants of the Renaissance in Europe.[9]

This chapter describes the innovations that created the Mongol fighting force and enabled the creation and administration of this revolutionary new empire. The chapter then explains the drivers of these innovations: great openness to diverse cultures, ideas and talent, and the empowerment of conquered peoples through rights, travel and trading privileges and a system of meritocracy unparalleled for its time.

Mongolian Innovations

Mongolian Horse: Basis for Innovation

In traditional armies of the time, the majority of warriors were foot soldiers. In contrast, the Mongol military consisted entirely of cavalry, with multiple horses per soldier.[10] Moreover, by the time of the Mongol invasions, Western and Middle Eastern horses were bred to be heavy and strong, to carry heavily armored knights in battle. In contrast, Mongolian horses bred and roamed in the wild rather than in captivity (see Figure 3.2). They had evolved to be short, sturdy and frugal. They could live off very little grass. They also had great endurance and could run for many miles without tiring.[11] The horses had evolved to adapt to the long, cold and harsh winter months from December through April. They grew a uniquely thick, tight coat that allowed them to withstand extremely cold temperatures. In addition, when snow got deep, the horses' sharp, strong hooves could break through snow and even ice to reach buried shoots of grass. On the Mongolian steppe, frozen for months at a time, the horses had adapted to using snow for daily water intake.[12] These adaptations of the Mongolian horses

Figure 3.2   Wild horses of Inner Mongolia

Source: Umut Rosa/Shutterstock.com.

provided the raw material for innovation. Training, strategy and tactics were still essential to create a fighting force that could conquer and build an empire.

Innovative Training of Horse and Rider

All the horses in the Mongol army had been tamed in the same way: lassoed from the semi-wild herd, and then ridden to exhaustion so that they accepted bearing a rider. The horseman would then befriend the animal, which would become part of the horseman's equine family. Once tamed, the horse assimilated quickly into the Mongol herd system and very rarely strayed, requiring minimal control from his rider to stay with the combat formation.

The Mongols innovated in their tack and harness. They were among the first cultures to employ stirrups.[13] Their sturdy stirrups allowed the rider to be more stably mounted and hence more accurate with his bow. The stirrups on a Mongol leather and light-wood saddle were very short, allowing the rider a high position, controlling the horse with his knees so that he could use both hands to shoot game with his bow, even turning around to shoot behind him. Training in practice fights and other formation drills was easier for the warrior because his mount responded quickly to abrupt changes in direction[14]—a crucial advantage in Mongolian-style swift equine warfare. Marco Polo noted of the Mongols, "Their horses are trained so perfectly that they will double hither and thither, just like a dog, in a way that is quite astonishing."[15]

Indeed, Genghis Khan's demands on his cavalry required that these horses had intensive training. The Chinese believed that this training gave the Mongolian horses great endurance. Having ridden their horses long distances daily, Mongol cavalrymen allowed the horses to eat when they were suitably calm.[16]

Because the Mongols disciplined their horses and learned to ride at an early age, they were able to concentrate on their archery skills with full use of both hands. When engaged in close combat, they also used the sword and wicker shield without having to grip the bridle.[17] Major Mongol units included two cavalries: the lighter riders who wore less armor, were incredibly swift, and were used for initial attack, and the heavy cavalry could move in for the kill.[18]

Mongols used their horses for a lot more than combat. The horses carried all that the Mongol warrior needed for his travels in their saddlebags—leather water bottles, cooking pots, dried goat and yak meat, yogurt and other foods made of cultured milk (also in leather pouches) and implements of all sorts.[19] Sometimes on long journeys, when food supplies were low and hunting was poor, warriors milked the mares of their herds for nutrition. When there was time, they brewed a mildly alcoholic beverage called *airag*[20] by fermenting the mare's milk; *airag* remains the national drink of Mongolia to this day. When things were at their worst, Mongols would nick a vein in their horses' necks and drink the blood. While they enjoyed horsemeat, they reserved it mostly for special occasions. Horses were worth far more to them alive.[21] Thus, Mongols could live off their own animals so that their armies needed little in the way of supply wagons. This frugality solved the challenge of traversing vast distances that other well-organized armies faced.[22] Relative to other armies of the times, Mongol armies traveled lightly, carried minimum supplies and even made heavy siege weapons on-site with recruited or transported engineers.[23]

The Mongols protected their horses in the same way they protected themselves. Before combat, they covered the horses with five-piece lamellar armor, which shielded every part of the horse's body, including the forehead, for which there was a specially crafted plate. Mongols used a lightweight wood and leather saddle rubbed with sheep fat to avoid cracks, which allowed for a long journey in relative comfort while retaining a firm seat for the cavalrymen.[24] Yet even with the horse and rider armored, the Mongol cavalries were very light compared to any others at the time, allowing them to execute tactics and maneuvers that were not possible for their more heavily armed enemies.[25]

The use of the Mongolian horse in warfare—combined with the extraordinary training of Mongol horsemen—made the great flexibility, speed and mobility of the Mongol army possible. Hence, in addition to the combination of training, discipline, intelligence and the constant adoption of new tactics, the swift warfare gave the Mongol army its big edge against the slower, heavier armies of rivals.[26] Because of how fast its warriors traveled, how quick it was in implementing and adapting strategy and tactics, and how agile and skilled it was in battle, the Mongol army outmatched most other armies it encountered in Asia, the Middle East and Eastern Europe.[27]

Continual Innovation in Strategy and Tactics

The astonishing military success of the Mongols baffled commentators of the time. They could not understand how a force of at most 100,000 lightly armored "barbarian" warriors, wielding small bows and arrows and spears, mounted on squat-looking ponies, seemed able to defeat almost every other army it fought against.[28] Typically, the Mongols' enemies considerably outnumbered them. For example, in the Battle of Ein Jalut, Muslim Mamluks had numerically much larger forces than the Mongols.[29] How could the Mongols continually triumph against such odds?

What European and Middle Eastern military leaders saw as weaknesses of the Mongol army were actually strengths. First, as already discussed, their small but strongly built horses were more agile than their large counterparts. Moreover, the lightweight laminated composite bow used by the Mongols had extraordinary range and power, rivaling that of the much larger and more cumbersome longbow or the compact but difficult-to-load crossbow—neither of which could then be efficiently used from horseback.[30] In addition, the superior arrows from a Mongol bow could penetrate plate armor at close range. The Mongol army was almost unstoppable with its unbreakable yet flexible chain of command, excellent communication and disciplined warriors. The Mongols used these advantages in mounts and weaponry to develop innovative methods of warfare.[31]

Speedy Battle Tactics

The Mongolian army could cover long distances even in frigid weather. Indeed, they used rivers that had frozen over as highways, on which they could travel swiftly to towns and cities, allowing them to attack when least expected, from a typically less defended side.[32] In fact, Mongol commanders viewed winter as the best time for war, while less hardy peoples took shelter from the cold. They used this strategy effectively in the attack on Russia.[33]

Each Mongolian cavalryman had his own four to six horses that he individually trained. The cavalryman would ride one for a while and then switch horses when one tired. This ensured that no horse was worn out and allowed the army to keep moving. This practice greatly enhanced the Mongol army's mobility over long distances.[34] Mongol warriors journeyed without carrying fodder for their horses. Instead, wherever possible, they traveled over grassy plains and along the banks of rivers, with good grazing and easy access to fresh water. These practices made the Mongol cavalry much swifter, more flexible and more maneuverable than the heavily built and armed European cavalry with one horse per man—not to mention infantrymen.[35] For example, during the Mongol conquest of Hungary early in 1241, the army could move close to a hundred miles in a single day.[36] No other army of the period, or centuries later, could match the Mongols' sheer speed.

The Mongol army was famous and feared for its cunning battle tactics. One of the Mongols' favorite strategies was a pretended retreat followed by a surprise attack, of which they employed two variants. In the first and simpler variant, a small force of cavalry would charge the enemy, then wheel abruptly and appear to flee, leading their pursuers into an ambush.[37] The second variant was for a larger cavalry force that had engaged the enemy to turn and withdraw as if they had been soundly defeated. The enemy troops, tricked into pursuing them, chased them for several days. For the Mongol warriors, patience was a virtue. They would bait their enemy for days or even weeks. They would pretend they were retreating and struggling, until they found a battlefield they liked.[38] Then, when the enemy was tired, loosely strung over a long distance, and at a disadvantage, fresh cavalry would emerge and attack, while light cavalry circled around the enemy to cut off retreat, while others would rain down arrows from the side. Then, with the enemy force in disarray, the Mongols' heavy cavalry would move in to attack with lances at close quarters.[39] Speed combined with cunning enabled these innovative strategies.

The Mongols used other deceptive tactics as well. For instance, they would tie leafy branches or large bundles of leaves behind their horses and drag them across bare ground, generating a dust cloud, usually on the far side of some higher ground from the enemy.[40] The goal was to generate fear by appearing to their opponents to be a much larger force than they actually were, thereby impelling the enemy's surrender to a delegation that would come to them under a flag of truce. Another trick was for the warriors to set prisoners or civilians astride their alternate horses and drive them along amid the ranks in advance of a battle in order to inflate their apparent numbers. Finally, when the army was encamped for a siege, soldiers built five fires at night and let the shadows of their horses grow large in the reflections of the flames.[41]

A brutal tactic used by the Mongols, which gave them their reputation, was to completely wipe out the populations of cities and towns whose leaders had refused to surrender, except for a few. These few were allowed to flee to the neighboring cities to tell of the Mongols' atrocities.[42] However, if a city or town chose to surrender, the people were spared and faced less severe consequences.[43] Surrendered cities had to pay tribute and support the Mongol army, but they remained largely intact.

Mobility and Communication

As the Mongol empire grew, it eventually incorporated almost 14 million square miles,[44] the largest contiguous land empire in world history. The administration and consolidation of this vast empire were achieved by good communications. Genghis Khan, highly innovative, constantly experimenting with new tactics, learning from enemies and adapting to new circumstances, built on the swift-rider messenger system he had devised to facilitate the rapid transmission of orders and information between army commanders. He developed a vast horse-borne communications network for his empire.[45] This system allowed commands to spread swiftly across the empire and news and reports to be brought swiftly to the capital, facilitating a timely response. The network involved placing postal/relay stations every 20 to 30 miles along key routes between strategic locations. The station consisted of a central facility for the riders, corrals for the horses and storage huts. There, a messenger rider would find shelter, a hot meal and well-rested, well-fed mounts. The messenger could then pass his message to the assigned next rider and rest for a day or two, or he might simply eat, take a fresh horse and keep going.[46]

Military Training and Discipline

The Mongol army was probably the world's most effective, disciplined and well-led fighting force until the mid-seventeenth century, when new centralized monarchies in Europe eliminated the feudal aristocracy and raised centralized armies. The European armies of the day had only a few trained professional soldiers that made up the relatively small cavalry: the knights and men-at-arms.[47] The rest were infantry conscripts: yeoman farmers, feudal serfs and artisans like blacksmiths. Artisans, as subjects of their lord, were required to fight in his army, but as a rule received only a few days of battle training. Among the Mongols, all males between the ages of 15 and 60 who could successfully complete the intensive, harsh training could be conscripted. To these men, serving in the imperial Mongol army was a supreme honor. The training and organization of that army replaced loyalty to the tribe with loyalty to the military unit, its commander and the khan.[48]

The simplest and yet most significant characteristic of the Mongol warrior was self-discipline. This training involved total commitment to military duty, unyielding loyalty to his unit and the army's chain of command, readiness to lay down his life for the success of the mission and contempt for the risks of an honorable death. The Mongol warrior was highly motivated because the reward for success in battle would be promotion as well as recognition from comrades and superiors. However, the penalty for incompetence or cowardice was execution. In general, Genghis Kahn rewarded warriors for performance and loyalty rather than kinship and distributed the loot from wars rather than hoarded it. Thus, warriors were intensely loyal, closely followed tactics prepared in advance and obeyed the orders of superior officers.[49]

When Genghis Khan came to power, he established an important standard of organization, discipline, equipment of the army and, most critically, the need to fight as a cohesive group. Genghis understood the crucial importance of making sure that every soldier's first loyalty was to his comrades and commanders, to the army as a whole, and of course to the khan.[50] As the Romans had before him, Genghis organized his army by the decimal system in groups of 10, 100, 1,000 and 10,000 with experienced and battle-tested commanders at each level.[51] Each unit could fight semiautonomously or in combination with other units in the field, as needed. As a rule, they could do this without continual supervision from higher in the chain of command because commanders at each level were trained to use initiative when faced with changing circumstances.[52]

Mongol military campaigns were prepared in advance using reconnaissance and espionage to gather needed information about the enemy territory and forces.[53] This intelligence gathering was followed by adept planning. The success, organization and mobility of the Mongol armies permitted them to fight on several fronts at once. They carefully thought out and rehearsed their military tactics. When commanders signaled, by firing flaming arrows into the air, striking drums or raising banners, warriors knew what they were supposed to do.[54] When not in battle, Mongol soldiers trained every day in horsemanship, bowmanship, hand-to-hand fighting, drills for coordination and battle formations. The commanders worked to anticipate every possible enemy tactic and then trained their soldiers to counter those tactics.

In sum, the Mongols' combination of extraordinary skill, tight discipline, inventive tactics and some of the most brilliant military commanders in history stunned all who fought against them. They lost very few battles—and when they lost, they usually returned to fight again another day, winning the second time around.[55] For over half a century from 1206 to 1260, Mongol armies were an unstoppable force that terrified the contemporary world.

Innovative Weaponry

Complementing their horse-based innovations in battlefield tactics, troop mobility and military communications, the Mongols made innovative use of weapons they developed or adapted from their conquered enemies.

Cavalry Weapons

The Mongol army's central weapon, on which many of its battle tactics depended almost as much as they did on the Mongol horse, was the unique Mongol bow. Unlike any other bow at the time, but like some modern sport hunting bows, the Mongol bow was made of laminated layers over a wooden core in a double-curved form. On the bow's inner (archer-facing) side was a thin layer of horn, which was resistant to being compressed as the bow was bent. On the outer face was an equally thin layer of sinew, which was resistant to being stretched. Together, this combination of properties allowed each draw of the bow to store more energy and impart it to the arrow when loosed.[56]

The construction of the Mongol bow made it both powerful and relatively efficient in terms of the relationship of force required to draw the bow to force of the arrow's release. This in turn meant that a version of the bow could be made small enough to be wielded effectively by a mounted rider. Moreover, riders were trained to time their arrow release to the exact moment the horse's hooves were off the ground to capture its momentum.[57]

The Mongol cavalryman carried multiple quivers stacked with arrows, strapped to their own backs and to their horses' cruppers.

As a rule, warriors were equipped with at least two bows. One of these was heavier and used while standing on the ground because of the force required to draw it; the lighter one was used to shoot while mounted. In pockets on their leather quivers were steel files that they used to hone their arrowheads, already hardened by heating them and plunging them in brine.[58]

As they used different bows for different purposes, Mongol warriors also employed various arrow types as well.[59] One type, used mainly for distance fighting, had an iron head and could travel as far as 200 meters. Another arrow type had a very sharp V-shaped tip intended to pass through the clothing and skin of enemies. A third type of arrow had holes in the carved bone head so that it would make a whistling sound in flight that could be used to tell others which direction to advance in. Another had a flammable head and was used as an incendiary weapon.[60]

The Mongol sword was a scimitar, which is a short sword with a slightly curved blade that widens away from the hilt to a very sharp, cornered straightedge rather than a point, a design first used in the Middle East. This design is another example of a traditional weapon adapted to the innovative Mongol style of warfare. Mongols used the scimitar for slashing attacks but also for cutting and thrusting, its size and shape making it easier to use from horseback than European long swords or broadswords.[61] The Mongol scimitar could be used with a one-handed or two-handed grip; the blade was usually about .75 meters in length, and the sword's overall length roughly one meter.[62]

Unlike the long, heavy pikes carried by medieval European foot soldiers, spears and lances were equally serviceable whether the warrior was mounted or on foot. They could be thrown like javelins to strike an enemy several meters away, or used in the hand for impaling enemies from the saddle.[63]

Siege Weapons

The Mongols would use almost any method to conquer a city, including diverting rivers and herding captives into the front line of battle. They also learned to use siege machines that hurled objects high and far. With these, they would either batter and destroy city walls or throw missiles, including incendiaries, over them. With the enemy's resistance lowered by bombardment and (if the siege was long) hunger, they would bring forward siege towers and swarm over the walls to attack.[64]

According to contemporary accounts, the Mongol army adopted these methods after they first encountered a walled city in northern Chinese territory. Commanders quickly realized that their cavalry was useless against these fortifications. Ever resilient, they recruited engineers from among their Chinese captives to build for them the catapults and siege towers needed to conquer more walled cities. From then on, these engineers traveled with the Mongol armies to rebuild siege engines wherever needed, using local materials where possible to avoid carrying the heavy wooden parts.[65]

The Mongols used siege engines in increasingly agile ways. For example, they learned how to make catapults that could be swiftly taken apart, carried in pieces on horseback and reassembled on site. They used these catapults with the technology of gunpowder, learned from the Chinese[66] (see Chapter 4). The Mongols would use their catapults to hurl gunpowder-based grenades. These weapons facilitated the Mongols' invasions of Hungary and Poland. Because the Mongols were always ready to acquire and use novel military technologies, their forces were not only the best organized, skilled and trained but also had at their disposal the most advanced weapons systems of their day.[67]

With their catapults, the Mongols launched a variety of objects over enemy fortifications, including incendiaries and heavy rocks.[68] Two more powerful weapons than the catapult were the *trebuchet* and the *mangonel*. Both these weapons could hurl heavy projectiles, usually boulders, at walls and fortifications.[69] Further, by recruiting highly skilled engineers from China, Persia and Europe, the Mongols played a decisive role in advancing the technology of guns and cannons prior to their diffusion to Europe (see Chapter 4).

Drivers of Innovation

The Mongols distinguished themselves from any other contemporaneous empire or kingdom with their high level of openness to cultures, ideas, skill and trade, and their empowerment of peoples through meritocracy and travel and trading privileges unequalled for its time.

Mongolian Openness to New Ideas, Trade, Peoples and Religions

Openness in the Mongol empire occurred through openness to diverse cultures, ideas and talent, unusual religious tolerance and the encouraging of trade.

Openness to Ideas

The Mongols were nomadic people who had relatively simple beliefs centering around Tengerist animism.[70] Unlike settled agrarian-urban peoples in neighboring regions, the Mongols, started no new major religions and created little in the way of dogma and literature. They practiced good animal husbandry but had no corps and produce agriculture or manufacturing systems.[71] So, the Mongols had little of their own to cling to and were highly receptive to new cultures, ideas and innovations, wherever they encountered them. Whatever they discovered in other cultures that seemed useful or appealing, the Mongols would unhesitatingly adopt for their own purposes and then spread throughout their territories.[72]

Mongols were eager to learn new technology (including printing), new religions and new forms of art.[73] At the time of the rise of Genghis Khan, Mongols could not weave their own cloth, cast the metal of their blades and arrowheads, make their own pottery, raise crops or bake bread. Yet, as their army conquered society after society, they picked up skills, which they later adapted to their own needs and spread around the conquered countries. This openness made them innovative technological and cultural cross-pollinators.

Openness to Religions and Talent

At the time of Genghis Khan, the Mongol people were Tengerists, which is a shamanistic belief system. *Tengerism* comes from a Mongolian word meaning "gods," notably those of nature and especially earth and sky.[74] However, once the Mongols began subjugating the lands around them, Genghis and his advisers did not impose their own spiritual beliefs and practiced religious tolerance. Instead of antagonizing conquered people by suppressing their religion, the Mongols allowed them to practice their faith freely, whether it was Buddhism, Taoism, Manichaeism, Islam, Judaism or Christianity. For example, the Mongol capital of Karakorum was probably the most open and tolerant city in the world at the time.[75]

Genghis Kahn mandated mutual toleration by followers of every faith in his domains—though he discouraged religious leader cults that might lead to revolt.[76] This was in sharp contrast to Europe of the time, where dogmatism prevailed and religious intolerance was common. To gain favor and loyalty with the conquered peoples, the Mongols exempted the leaders of all faith communities from taxation. The empire's vast territories, which necessarily included followers of all these religions and more, would have been difficult to govern effectively if the khans had not pursued this policy. Even when they themselves became adherents of one of these foreign religions, the Great Khans and minor khans maintained and enforced this policy evenhandedly and rigorously. For instance, Kublai Khan was a Tibetan-style Buddhist, but he permitted peoples he ruled to practice their own religions.[77]

This policy of tolerance extended throughout the ranks of society from bottom to top. Genghis Khan and his descendants employed Christian, Buddhists and Muslims as engineers and senior civil servants in the government of the empire. Administrative talent was the primary qualification for advancement in civil service, just as military talent was the primary criterion for promotion in the army, and engineering expertise for construction and manufacturing in the empire. Even some of Genghis's closest advisers were followers of other religions. To the Mongols, in fact, religious tolerance was not only wise imperial policy but also an integral part of their basic attitude to life—which included curiosity about how other people viewed the world.[78] When Ogedai Khan created the Mongols' imperial capital, Karakorum, he permitted religious leaders to build temples, lamaseries, mosques and churches for their worshippers.[79]

The Mongols were also open to talent from all over the empire. They needed decisive rulers, skilled politicians and writers to draft and propagate provincial laws and decrees and administer the laws and decrees of the khan, send in reliable reports, maintain accounts and record the resolution of internal conflicts and the dispensation of justice. So they were open to and recruited talent from all over the empire. Along with engineers for siege weapons, the Mongols recruited anyone who was useful to them, such as artists, musicians and especially administrators. Mongols were not much for administrative and clerical tasks like keeping records, so they found people who were good at it and moved them around the empire.

This openness to other belief systems, religions and talent allowed the Mongols to harness the intellectual energy of their subjects for innovation and enterprise throughout the empire.

Encouraging Trade, Ideas and Cultural Exchange

In terms of the empire's lasting impact, the Mongols were big supporters of merchants and trade. They were eager to foster exchange of every kind among the cultures under imperial sway.[80] Part of this attitude was driven by background. The Mongols were primarily herdsmen. They had little agriculture or manufacturing of their own. Thus, from the beginning, they relied on trade. This background made them very open to trade. As the empire grew, the Mongols traded on an unprecedented scale.[81] Just for his army, Genghis Khan imported iron arrowheads and spearheads, steel swords for war, metal parts for the horses' tack, leather for armor and clothing.

As a result, for the hundred years of the empire's ascendancy, international trade flourished as never before. Tea, rare and valuable spices, bolts of silk and artworks such as scroll paintings traveled from China to Europe and the Middle East. Arab medical manuscripts, works on astronomy and alchemy, gold coins and jewelry, and fine porcelain traveled east to Asia. Along with such expensive goods, ideas and technologies also moved both ways along the Silk Road.[82]

Genghis offered his imperial protection to the merchants who came from the East and West. He also awarded them a much higher status than the social rank assigned to them by the Chinese, who did not think highly of trade and traders, or the Persians.[83] Merchants received not only protection and high status but also financial help in the form of tax exemptions, low-interest loans and other aid from the ruling khans. Over time, the loose bundle of mostly partial trade routes between East and West across northeastern Europe and Central Asia were linked together by travelers and became the Silk Road, a continuous network of land routes between Europe and Asia that became legendary.[84]

All traders and diplomats who could show proper documentation and authorization were under the khan's protection while traveling through his territories. Genghis Khan, in fact, offered traders a form of imperial passport that gave them safe passage along the Silk Road.[85] This innovation, among others, contributed to the rapid expansion of overland trade.

Even then, merchant caravans crossing wide, largely uninhabited areas faced not only natural disasters like sandstorms and floods but also attacks by bandits and renegades. To survive these threats, they needed to be in large caravans, with as many as 70 to 100 men including trained mercenary warriors for protection. This setup made caravans expensive. The loss of one such caravan could completely wipe out a merchant. Knowing this, the Mongols devised an association, the *ortogh*, whereby member merchants would pool their capital to share both the expenses and profits of a caravan—a precursor of the state-chartered mercantile companies of the seventeenth and eighteenth centuries. Instead of high tax rates that prevailed previously, the Mongols gave traders tax exemptions. For convenience, the Mongols used paper currency, an innovation adapted from the Chinese, backed by silk and precious metals.[86]

Because the Mongols had both opened up (and significantly secured) the trade routes, and because they did so much to support and encourage long-distance trade, they also facilitated major knowledge transfers—technical, scientific, mathematical, philosophical, artistic—between Christian Europe, Persia, the Muslim world, China, Korea and Japan. By the same token, they facilitated the formation of Eurasia as a single vast commercially integrated region. In particular, Eastern goods and innovations that reached Europe most likely stimulated the emergence of the

Renaissance in Europe.[87] This is perhaps the greatest legacy of the Mongol empire.

Mongolian Empowerment through Integration and Meritocracy

Integrating the Conquered into Mongol Society

Following each new conquest of a nation or region, the Mongols removed the governing elite—typically the top layer of an aristocracy—and trained and elevated a new provincial administrative leadership drawn mainly from the educated class, such as clerics or mandarins.[88] When nomadic tribesmen such as those in Georgia, Armenia, Turkey and elsewhere in western and central Asia surrendered to the Mongols, they were recruited and integrated into the imperial army. Consequently, each time the empire conquered and incorporated lands farther and farther away from Mongolia, the army continued to grow as non-Mongol warriors volunteered or were pressed into service.[89]

Creating a Mongol Identity

Before he could begin to build his empire, Genghis Khan first had to unify and rally all the tribes of Mongolia, which up to that time had often been in conflict. In fact, it was only after he had achieved this aim that he received the title of Genghis Khan, which means "Supreme Ruler."[90] In so doing, he laid the foundation for the development of a unitary Mongol national culture and sense of belonging. This included the creation and display of symbols, like the *tug* banners with their horsehair streamers and golden crests that to this day represent the unity of the Mongol people.[91] But in a way typical of his organizational and strategic genius, Genghis Khan took more practical steps as well. In his army, he mingled fighters from different tribes in order to minimize any mutually reinforced tribal loyalties and replace them with loyalty to Mongolia and himself as Great Khan. The Mongolian people today, as a rule, remain proud of the empire and the feeling of national identity and achievement it gave them.[92]

Meritocracy

Genghis Khan organized, trained and equipped a cohesive, flexible fighting force capable of independent operations. Both commanders and common soldiers were as a rule given wide discretion in how they carried out their general orders, so long as the plan's larger and longer-term objectives were met and orders from above obeyed without delay. In this way, the Mongols obviated the problems associated with too-rigid command chains, which have bedeviled armies throughout history. But this flexibility was made possible by the strictly enforced requirement that soldiers be unconditionally loyal to their unit comrades, their commanders all the way up the line, and above all to the khan.[93]

Although ultimate command was always in the hands of the Supreme Khan, Genghis Khan introduced and enforced, as did later khans, the rule that individuals would be promoted to positions of greater responsibility and authority solely based on demonstrated ability.[94] This rule resulted in an army of unrivaled excellence, all the way from common troopers to top commanders. Virtually any Mongol soldier was superior by a wide margin in training, discipline and fighting skill to his Western counterparts. This high quality through the ranks together with the system of merit-based promotion reciprocally ensured the commanders' competence and integrity. This system empowered leaders at each level—from the decade to the army of 10,000—and thus trusted them with a great deal of independence in achieving their assigned military objectives.[95] It also fueled competition among individuals who realized that they could rise to a position of power if they developed their skills.

As both enforcers and beneficiaries of this system of merit-based promotions, Mongol commanders functioned as leaders in an organizational context that allowed initiative at each level under minimal direct central command.[96] So, leaders were able to gain optimum effect from the unmatched swiftness and skill of their cavalry and the accuracy and force of their weaponry.[97] However, the sheer speed of Mongol cavalry made continually updated information from across the entire combat zone absolutely essential. As a rule, therefore, by means of continual scouting forays and regular long-range arrow messages, Mongol commanders had a much clearer and more current sense of conditions across the battlefield at any given moment than did their opposite numbers on the enemy side.[98] When this system was used improperly, Mongol commanders experienced major defeats. For example, during the Persian campaign at the battles of Fergana Valley and Parwan, the scout system was unable to convey to top commanders the up-to-the-minute information that their swift equine warfare tactics necessitated, and the Mongol army lost both battles.[99] Despite temporary setbacks and very few defeats, as one campaign after another spread out across Europe and Asia, the Mongol high command developed into a cadre of highly empowered, superb strategists and tacticians, whose performance has seldom been matched by any other military leadership before or since.

Genghis Khan opposed self-perpetuating aristocracies and did everything he could to discourage them (including executing those of conquered nations wholesale).[100] For example, early on when his soldiers looted a captured town, Genghis mandated that the haul be shared among the troops and not go to tribal aristocrats, as had been the tradition.[101] More importantly, he promoted those beneath him according to their abilities, courage and loyalty to the khanate—a principle he first applied to his armies and later to his government and civil service. In the army, if a soldier demonstrated bravery in battle and cleverness in tactics, he would undergo rapid elevation. If, however, a general lost a battle, he would be stripped of his rank and broken back to common trooper. With the same thoroughness and concern for the long-term welfare of his people, Genghis assembled a legal code known as the Yasa,[102] which went on to influence the law in many other states over the next seven centuries. The Yasa mandated mutual love and respect among all people, advocating sharing food with the hungry and caring for the weak and the elderly. It also gave equal protection to all under law as well as complete religious tolerance. By requiring that equal opportunity for advancement be afforded to all, whether rich or poor, it also reinforced the rule of promotion by merit.[103] In this way, the Mongols created stability in their realms by empowering their diverse subjects. They did so by using Genghis Khan's legal code, which gave common people in most cases more rights than they had possessed under their former rulers.[104]

Decline of the Mongol Empire

In the late thirteenth century, the Mongol empire split into four smaller khanates or subempires: the Yuan empire of China that Kublai Khan created, the khaganate that dominated Central Asia, the ilkhanate that ruled the Middle East and the Russian-centered Golden Horde.[105] The empire probably reached its peak on the rule of Kublai Khan, the grandson of Genghis Khan, who conquered China. At this time, the empire encompassed much of northern Asia from China in the east to the Danube in Hungary, including parts of the Middle East. While parts of these territories were devastated during conquest, the well-run and efficient Mongol administration made up for the damage over time and often made improvements over previous social and economic conditions. Agriculture, craft and manufacturing flourished, and trade greatly expanded across the entire immense domain.[106]

In the span of less than two centuries, the Mongol empire arose, peaked to its largest size and then dissolved. Several factors may be responsible for its decline.

First, the Mongols were a small people numbering about a million. They ruled a vast and diverse empire that was about 12 million square miles in territory and numbered hundreds of millions in population. Even with a swift communication system based on the Mongolian horse and communication posts, the empire was practically too big to administer with any degree of control and standardization. In this respect, the split into smaller khanates was inevitable, and yet practical.

Second, the empire was unable to sustain its openness over time. From the mid-thirteenth to the mid-fourteenth centuries, the empire experienced relative peace resulting from the stabilizing effect of united rule over vast parts of Eurasia that facilitated communications, commerce and transfer of skill and knowledge across huge landmasses. However, unlike their predecessors, which were sedentary empires, the nomadic society of the Mongols had no strong traditions of succession. A few generations after Genghis

Kahn, the empire splintered, with a struggle for succession among his descendants. The division into khanates blocked skill and knowledge transfer from one end of the united empire to the other. The Silk Road traversed too many diverse regions and khanates to be perennially effective. It became prone to theft, pirates and taxes from the nations through which it passed. Moreover, it also became a pathway for the spread of disease, especially the plague, which created havoc in Asia and Europe. Genghis Khan's noble values of openness to diverse peoples, empowerment of oppressed subjects, and encouragement of trade, were tough to maintain in the face of numerous countervailing forces across the different ethnic groups, religions, and rulers of the vast empire. Genghis Khan's philosophy did not last more than a few generations past his death. In addition to the decline in openness, the most prominent driver of Mongol innovation, which was empowerment through meritocracy, similarly declined.

Third, swift equine warfare greatly diminished with the rise of a new technology, gunpowder. Gunpowder initially developed in China and then was advanced by the Mongols. Through them it spread quickly to the Islamic world and Europe, where it advanced greatly, well beyond its origins in China and its development by the Mongols (see Chapter 4). This new technology of gunpowder favored an infantry over cavalry, rendering the traditional Mongol warfare less effective (see Chapter 4). These processes occurred at the same time as other powers, especially in Europe, surpassed the Mongols in the level of competition between states and their empowerment of entrepreneurs and innovators (see Chapter 4).

Thus, internal power struggles led to division into khanates, which in turn led to a decline in openness, empowerment and innovation. These changes ultimately led to the decline of the Mongol empire. Various successor empires in China and northern, southern and eastern Asia claimed descendancy from the Mongols for centuries thereafter. However, aside from their rulers' remote family connections to Genghis Khan, they were not true successor nations of the Mongol empire in terms of its philosophy.

4

How Gunpowder Shaped the Fortunes of Nations

In 1500, 1521 and 1522, the small nation of Portugal attacked the mighty Chinese empire thousands of miles away from Portugal. The Chinese repulsed those attacks. However, when the British attacked a couple of centuries later, the Chinese were less successful. In all these wars, gunpowder weapons were a key driver of success. Ironically, gunpowder technology was first invented in China, where its early development occurred over a period of about 400 years before it diffused to Europe.

Why were the Chinese, who invented gunpowder, not the dominant power in firearms between 1450 and 1800? When, how and why did the Europeans grow superior to the Chinese in gunpowder technology? What drove the Chinese, the Ottoman Empire and various European nations to innovate in gunpowder technology? This chapter seeks to address these issues.

Gunpowder is an important innovation that profoundly impacted the rise of nations. Specifically, nations that adopted the most up-to-date gunpowder innovations acquired an edge in military warfare and overcame competing nations. Those that lagged by even one technological generation suffered immensely if competing nations were ahead of them in the technology.

Gunpowder is a mixture of sulfur, charcoal and potassium nitrate (also known as saltpeter). While saltpeter is the oxidizer in this mixture, sulfur and charcoal act as fuels.[1] In warfare, gunpowder served three primary purposes. First, gunpowder acted as an incendiary. It was often mixed with other inflammatory

Figure 4.1　Approximate timeline of inventions in gunpowder weapons in ancient China

First Cited	Invention	Dynasty
904	Gunpowder arrows	Song
970	Improved version of gunpowder arrows	Song
1000	Gunpowder pots and gunpowder caltrops	Song
1002	Fireballs	Song
1044	Thunderbolt bomb	Song
1126	Thunderclap bomb (incendiary)	Song
1127	Fire lance known as ancestor of the gun	Jin
1130s	Gunpowder firecrackers	Song
1207	Thunderclap bomb (explosive)	Song
1221	Iron fire bomb	Jin

Source: Dates compiled from T. Andrade, *The Gunpowder Age*, 2016.

materials and the ignited mixture hurled to set distant objects on fire. Second, gunpowder was used as an explosive that either damaged structures, injured military personnel or terrified enemies. For this purpose, often stone, metal or porcelain pieces were incorporated into gunpowder weapons to aggravate the damage caused when the weapon exploded. Third, gunpowder was a strong propellant that sent pellets, stone or metal balls, or bombs toward enemy targets. The purpose of gunpowder determined the proportion of potassium nitrate.[2]

Early innovations in gunpowder from the mid-ninth century to the beginning of the twelfth century focused on the creation of a more controllable and effective incendiary mixture than any other extant fire-producing materials like oils (see Figure 4.1).[3] Next, innovations developed various delivery mechanisms to project the incendiary mixture toward target objects or enemies. These innovations included gunpowder arrows, animal-driven incendiaries and fire lances. Later innovations focused on developing explosives, such as bombs, which could damage structures as well as terrify and injure troops.[4] Subsequent innovations harnessed the propellant force of gunpowder to bombard enemy structures and personnel.[5] These innovations resulted in guns, cannons and rockets.

This chapter traces the history of gunpowder—an innovation that provided nations a critical edge in warfare. It then explains what factors drove its development and diffusion, especially competition among nations and openness to external technological developments.

Innovations in Ancient China

Intense competition together with openness to foreign ideas and incentives for engineers led to the development of four early Chinese innovations in gunpowder weapons: gunpowder arrows, animal driven incendiaries, fire lances and bombs.

The Chinese discovered saltpeter by the mid-first century CE and sulfur in the second century CE.[6] In 142 CE, a Chinese alchemist wrote about a mixture of three powders (possibly gunpowder) that would "fly and dance" violently.[7] By 492 CE, Chinese alchemists discovered the burning property of saltpeter. Over the centuries, Chinese alchemists experimented with various proportions of saltpeter, sulfur and charcoal to create a potion for immortality. In the middle of the ninth century, alchemists developed an effective formula that had enough reactivity to produce flames. Researchers attribute this development to the innovation of gunpowder.[8]

Gunpowder Arrows

Gunpowder arrows were probably the first gunpowder weapons used in war in 904 CE in a battle fought at the end of the Tang dynasty (618–907 CE).[9] Gunpowder arrows consisted of gourd-shaped combustible materials, including gunpowder, mounted on an arrow. The fuse attached to the mass was lit before releasing the arrow toward a target. Gunpowder arrows were an improvement over fire arrows used in warfare in China at least from the third century CE.[10] Early gunpowder could burn only when exposed to oxygen. Gunpowder arrows rushing through air provided enough oxygen to burn the gunpowder.[11] Further innovations over the next hundred years led to widespread use of these weapons in wars.

Figure 4.2   Map of Song and its rivals

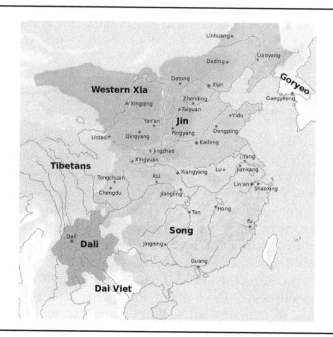

Source: Mozzan/Wikimedia Commons.

Animal-Driven Incendiaries

Animal-driven incendiaries were probably the next big development in gunpowder weapons. The advanced Song dynasty (960–1279 CE) in China was frequently under threat from its neighbors (see Figure 4.2).[12] This intense competition motivated the Song to experiment with gunpowder weapons to gain a competitive advantage. Accordingly, the Song accorded high priority to the research and development of gunpowder weapons.[13] For example, the firebird was a bird that carried a mixture of gunpowder tied to its neck to set fire to enemy formations.[14] Likewise, fire beasts were animals whose rumps were attached to a mixture of gunpowder, ignited and driven to enemy sites.[15] Sometimes, spears were tied to the rushing animal to injure enemy warriors. The animal confused and scared enemies, and set their objects on fire.

Bombs

Bombs and other explosive materials were the next big innovation in gunpowder that occurred in the twelfth century, when the Song faced off against its strongest competitor until then: the Jin. The intense competition between the Song and the Jin over the next century triggered the development of various powerful gunpowder weapons. Under pressure of such rivals, the Song incentivized alchemists to experiment with gunpowder recipes.[16] With the gradual increase in nitrate levels, gunpowder became ever more explosive as opposed to being primarily incendiary in the past. The explosion had two effects apart from setting things on fire—scaring and physically harming enemies. Early explosives most likely served only the first purpose. After about half a century of such use, gunpowder weapons became powerful enough to cause bodily harm. Possibly the first known explosive weapon, the thunderclap bomb, was used in the Song-Jin battle at Kaifeng in 1126. The thunderclap bomb contained a bamboo tube that was packed with gunpowder and porcelain pieces. These contents were wrapped into a ball consisting of paper, and the ball was again coated with gunpowder. The purpose of bamboo was to accentuate the explosive effect and scare the enemy. The bombs impressed the Jin, who retreated.

Because of this fierce competition, by the early thirteenth century, the Jin developed a powerful bomb called the iron bomb or thunder-crash bomb.[17] Compared to the thunderclap and leather bombs, in which the explosive contents were enclosed in weak casings made of paper or leather, the iron bomb had a strong casing made of cast iron. The Jin probably used the iron bomb for the first time when they attacked the Song city Quizhou in 1221.[18] The bombs were powerful enough to injure humans. In 1232, the Jin used a deadlier explosive called the heaven-shaking thunder-crash bomb against the invading Mongols. Most probably, this bomb had a higher nitrate content than the thunder-crash bomb.[19]

Fire Lances

The next big innovation in gunpowder technology, the fire lance, also emerged from the Song-Jin rivalry. After defeat by the Song, the Jin used captive Song artisans to study and build more powerful gunpowder bombs. The Jin returned a few months later to Kaifeng and used gunpowder arrows and bombs to defeat the Song in 1127. The Song fled southward and eventually established a new capital in Hangzhou. The Jin chased them, and in subsequent intense competitive battles, a new weapon that would change the course of human history emerged: the fire lance, the forerunner of the gun. The Chinese had discovered that when heated, the air trapped in between the inner and outer walls of the bamboo stalk expanded to produce a cracking sound. Ancient Chinese used fire crackers contained in bamboo to ward off evil spirits even before the invention of gunpowder weapons.[20] This knowledge may have triggered the enclosing of gunpowder in a container to ignite an explosion. A fire lance is a long staff at the end of which is a tube filled with gunpowder. When gunpowder is lit, flames spew forth from the tube toward the enemy.[21] Various historic sources indicate that the Song dynasty invented the fire lance,[22] which may have been first used in 1132, when the Song defended its city against the Jin.

Thus, intense competition in China led to major developments in gunpowder weapons including gunpowder arrows, animal-driven incendiaries, fire lancers and bombs.

Innovations under the Mongols

The Jin's success with explosive gunpowder weapons did not last long. The invading Mongols soon turned out to be the Jin's deadly competitors. The Mongols' encounters with the Jin profoundly changed the nature of Mongol warfare. Beforehand, the Mongols predominantly depended on their swift equine warfare to defeat their competition (see Chapter 3). But when the Mongols faced walled cities protected with catapults and gunpowder weapons during their invasions in North China in the early twelfth century, they looked for artillery options. Ambughai, the head of the artillery division under Genghis Khan, trained catapult operators, bomb throwers and bowmen who would deploy gunpowder weapons in future conquests against the Jin.[23]

In 1232, Ogedei Khan, son of Genghis Khan, who founded the Mongol empire, plotted the capture of the Jin capital,

Kaifeng. Gunpowder weapons played a decisive role in the battle of Kaifeng in 1232.[24] The Mongol military commanders were open to innovations and drafted Chinese experts into their forces. The latter transferred gunpowder technology to the Mongols. Once they learned about gunpowder, the Mongols advanced gunpowder weapons through their own innovations. When the Mongols attacked Kaifeng with gunpowder bombs, the Jin replied with heaven-shaking thunder bombs and fire lances. But the Mongols had encountered these weapons earlier in 1231 and were better prepared this time. After a year-long bloody war that killed hundreds of thousands of men, the Mongols conquered Kaifeng in 1233, ending the Jin dynasty. Through the adoption of and innovations in gunpowder, the Mongols defeated their own teachers of the technology, the Jin.[25]

After the victory against the Jin, the Mongols attempted to conquer the Song, who had powerful gunpowder weapons of their own.[26] When Kublai Khan took control over the Mongol forces in 1259, he engaged in intense competitive battles with the Song. Both sides used gunpowder weapons, but neither gained the upper hand. At the time, the weapons of choice included gunpowder arrows, thunder crash bombs, iron bombs and fire lances. The Mongols added to this arsenal by developing trebuchets, handguns and cannons.

Trebuchet

Eventually, Kublai Khan received assistance from the Mongol Ilkhanate in Persia. He drafted engineers from Persia and Syria to help build a new trebuchet, which worked on the principle of counterweight.[27] These trebuchets came to be known as Muslim catapults. In these new catapults, a heavy load at one end of the hurling lever provided the required mechanical power to hurl a very heavy projectile. These new catapults had a range of about 500 meters and could launch projectiles weighing over 300 kilograms. The Muslim catapult was successfully deployed in 1271 against Xiangyang, who eventually surrendered in 1273. The Mongols' openness to foreign experts played a critical role in this victory. After continuous efforts over half a century, the Mongols finally defeated their fierce competition—the Song—and established the Yuan dynasty (1271–1368) in China. In the course of battles, the Mongols developed three major innovations in China, guns, handguns and the cannon.

Gun and Handgun

A first prototype of the handgun is likely to have been developed during the Mongol rule of China based on fire lances. Through years of battles, their military technicians discovered that flames emitted from a tube could carry along some solid objects with reasonable force. So, they added metal objects, porcelain pieces and pellets to gunpowder to hurt the enemy. Initially the tubes were made of bamboo.[28] As gunpowder increased in power, innovators felt the need to have a stronger barrel than the bamboo ones. The Mongols in China were the first to use a metal barrel in a fire tube to contain the propellant. This innovation was a major turning point in the evolution of guns. The *Fire-Dragon* manual mentions the "bandit-striking penetrating gun" used by the Mongols, a three-foot-long barrel made of iron and operated by foot soldiers.[29]

Historical records indicate that Li Thing, a Yuan military commander, used handguns to defeat Nayan, a Christian prince who revolted against Kublai Khan.[30] This device was a significant departure from bulky catapults and trebuchets used until then. A nondated bronze gun, weighing about 3.5 kilograms and over one foot long, was found near the site where Li Thing suppressed Nayan's rebellion in 1288. It seems that the Mongol Yuan dynasty centered in China completed the transition from bamboo-barreled firearm to metal-barreled firearm.[31] The Xanadu gun, built in 1298 by the Mongols, is the oldest dated gun. The inscription of serial number and manufacturing information on the Xanadu gun indicates that gun manufacturing was probably quite systematic during the time of the Mongols.[32]

Cannon

During the thirteenth century, the Mongols under the Yuan dynasty also developed larger fire lances termed "eruptors," the predecessors of the cannon.[33] One such weapon was the "multiple bullets magazine eruptor." The barrel of the eruptor was made of cast bronze and mounted on a wooden carriage. The barrel was filled with gunpowder and hundreds of lead balls. The eruptor could be rotated in all directions. The lead balls deterred enemies on the battlefield. The gunpowder used in this eruptor most likely had low nitrate content, so the flames pushed the lead balls out of the barrel with considerable force. The "flying-cloud thunderclap eruptor" was a weapon that used a mixture of gunpowder and cast iron explosive bombs. These explosives detonated on landing, causing fires and casualties. The "poison-fog magic-smoke eruptor" used an explosive shell containing gunpowder and poisonous materials, which could produce smoke and hurt the eyes and faces of enemy soldiers upon explosion. In all these cases, the projectiles did not close off the barrel of the weapon, a later Mongol development.[34]

Thus, the Mongols' adoption of gunpowder led to a new era in the development of this technology, especially development of the gun, handgun and the cannon. Most importantly, they spread the technology to the Middle East and Europe through their military expeditions, their support of open trade and their exchange of ideas.

Innovations in the Ming Dynasty

Emperor Zhu Yuanzhang, also known as the Hongwu emperor and founder of the Ming dynasty, defeated rivals and the Mongols through the use of gunpowder weapons. His competitiveness in terms of territorial achievements fueled great innovation in gunpowder technology and guns, which played an important role in his success. This competition was a major driver in the development of artillery and a significant aspect of the Ming military. The Hongwu emperor established an industry of gunpowder weapons. By 1380, gunners comprised 10 percent of the Ming army.[35] By 1466 this proportion increased to 30 percent.

To compete with its new rivals, the Ming dynasty developed various new cannons.[36] One such cannon, the crouching tiger cannon, weighed over 20 kilograms and was two feet long. This cannon was pinned down to the ground to mitigate the recoil.

The cannon barrel could hold 100 pellets. A long-range cannon was much larger, weighing 72 kilograms and being two feet, eight inches long. It could fire a single lead ball weighing 1.2 kilograms or 100 small lead pellets. The double bowl-mouthed cannon had two guns, pointing in opposite directions, set on a horizontally rotating pivot. This gun was an innovative solution for the delay in successive shots. While a soldier was loading the first gun after a shot, the other gun could be rotated toward the target for a successive shot. The great general gun was three feet long and weighed about 90 kilograms, transported on a four-wheeled carriage. Advanced metallurgy during the Ming dynasty helped make the barrel straight, with ribs or rings to strengthen it.

It took five centuries after its invention in China for gunpowder and gunpowder weapons to arrive in Europe in the thirteenth century.

Diffusion to Europe

Several theories exist about the diffusion of gunpowder from China to the Middle East, Europe and rest of the world.[37]

The first one suggests that European travelers who visited the Mongol court in the mid-thirteenth century learned about gunpowder. William of Rubruck, a French friar sent by King Louis IX of France on a mission to the Mongol court, was one such traveler. He spent four years in the Mongol capital, Karakorum, and wrote a comprehensive account of his stay there. He may have introduced gunpowder to his alchemist friend Roger Bacon, who produced the first account in Europe on gunpowder in his work *Epistola de secretis operibus artiis et naturae* in 1267.[38]

The second theory is that foreign military technicians who worked for the Yuan dynasty transmitted knowledge about gunpowder back to their home country. Kublai Khan preferred to use foreigners because he did not always trust local Chinese people to run his administration. So, men from Persia, Syria and other Arab lands were recruited as technicians in military service. These technicians may have conveyed gunpowder technology to the Middle East.[39]

A third theory suggests that the Mongols brought gunpowder to Europe during their thirteenth-century invasion. In 1219,

Genghis Khan led the Mongol invasion of Transoxania, a region that corresponds to modern-day Uzbekistan, Tajikistan, southern Kyrgystan and southwest Kazakhstan. The Mongols had an army with catapult specialists, who may have used gunpowder bombs. Mongols used these weapons again in Transoxania in 1220 and later in sieges in southwest Russia in 1239–40. These military expeditions may have transmitted knowledge about gunpowder to Europe.[40]

A fourth theory suggests that the technology diffused through European merchants. For example, a colony of Italian merchants traded in the Ilkhanate in Persia, the southwestern region of the Mongol empire. Some of the prominent merchants in this colony, in their diplomatic roles, closely interacted with Italian city-states, the pope and lords in England in the late thirteenth century to build a Mongol-Christian alliance against the Muslims.[41] These merchants may have spread the knowledge about gunpowder to Europe.

By the 1320s, guns were available in various parts of Europe. A book written in 1326 by an English scholar[42] contains an illustration of a cannon, which is the first written evidence of existing firearms in Europe.[43] In the same year, the government of Florence, Italy, ordered officials to make guns to defend the city.[44] These guns are remarkably in line with Chinese design.[45] In 1327, the Italian city of Turin paid technicians for making guns. In 1331, the Germans used guns during the siege of Cividale, Italy.[46]

Early European Developments

Due to intense competition among European countries, firearms quickly diffused and advanced in Europe.[47] Importantly, Christian Europe had a strong tradition of metallurgy owed to the technology of casting large metal church bells. The technological gap from casting bells to casting cannons was rather minor. By 1341, the French town Lille possessed guns.[48] In 1344, the German towns Ehrenfel and Mainz had firearms.[49] In 1346, Aix-la-Chapelle in Germany had wrought iron cannons. By 1348, the Dutch town Deventer possessed cannons.[50] The English had considerable success using firearms in warfare, especially in various battles against the French during the Hundred Years' War (1337–1453).

For instance, guns played an important role in England's triumph in the Battle of Crecy in 1346 against the French.[51] Later, in 1356, the English used guns to successfully defend the Breteuil castle against the French who tried to recapture the castle.

Early European guns were small. The Loshult gun, one of the earliest European guns, dated to the late fourteenth century, weighed nine kilograms and was only 30 centimeters long. This gun was similar to the Mongol's Xanada gun of 1298 that weighed six kilograms and was 35 centimeters long. In the early thirteenth century, European guns were mostly used as antipersonnel weapons or to burn wooden structures. Philip the Bold (1342–1404), the Duke of Burgundy, meaningfully influenced the evolution of guns in Europe. Driven by his keen interest in firearms, he built one of the most effective artillery armies in Europe and converted Burgundy into one of the most powerful states in Europe using gunpowder weapons.[52] He demonstrated the power of big guns in 1377 during the siege of Odruik in France. The castle in Odruik, held by the English, had fairly strong walls, which the French had earlier tried in vain to breach using small guns. Philip, who was France's ally against the English, used several large guns that could fire projectiles that weighed about 90 kilograms.[53] The guns easily breached the castle walls, and Odruik capitulated. Thus, for the first time ever in the history of gunpowder weapons, guns became powerful enough to destroy strong walls. This siege marked the beginning of a new era in European firearms.

The constant competition among nations in Europe triggered innovation in military technology[54] and a race to build larger and more powerful artillery. One primary example was the Hundred Years' War between England and France. By the early 1400s, England and France built large guns that could batter castle walls. Henry V, who at age 26 became the king of England in 1413, led the English in the race to build bigger guns. When negotiations with France failed, he invaded France in Harfleur and demolished the castle walls with 12 large guns.[55] Though the French used guns to defend Harfleur, their weapons could not counter the English guns. With superior guns, Henry V conquered city after city until Charles VI finally surrendered in 1420.

Later, in 1429, French forces defeated English forces in one of the most important gunpowder battles in European history.[56] Led by Joan of Arc, the French used cannons in various subsequent battles to recapture territories lost to the English.[57] Joan of Arc's victories shifted the momentum in France's favor. In 1437, Charles VII recaptured Paris, the capital he was driven away from 19 years earlier. He substantially reorganized the army and built a strong artillery in France. Eventually, with the help of superior artillery, France drove the English out and brought an end to the Hundred Years' War.[58]

Over the course of the fourteenth and fifteenth centuries, European guns grew very big. The Burgundians built a gun that could throw projectiles weighing 200 kilograms.[59] In 1431, the Dutch built a cannon called the Dulle Griet that weighed 12,000 kilograms and could fire projectiles of 300 kilograms. In 1411, the Germans built a gun called the Faule Mette (Lazy Mette) that could shoot stones weighing 400 kilograms.[60] During this period, other Western European nations like Portugal and Spain acquired gunpowder weapon technology.

However, despite this growing European progress in gunpowder technology, a new power was ascending. The forces of openness and competition in the Middle East drove the development of gunpowder technology that rivaled that in Europe.[61]

Developments in the Muslim World

Prior to the establishment of the Yuan dynasty in China, the Mongols attacked various parts of Central Asia, Iran, Iraq and Syria in the mid-thirteenth century. The Mongols used gunpowder weapons in their campaigns in West Asia from 1253 onward. These campaigns were led by Hulagu Khan, grandson of Genghis Khan and brother of Kublai Khan. The campaigns resulted in the establishment of the Ilkhanate of Persia. The earliest recorded instance of gunpowder weapons here was during the attack on Syria in 1256.[62] Sometime in the last quarter of the thirteenth century, a Syrian alchemist wrote a military manuscript that contained gunpowder recipes and purification techniques of saltpeter.[63] The recipes and techniques in this book resemble gunpowder-related information in various Chinese texts before that time,[64] suggesting that gunpowder technology had arrived in the Islamic world from China.

The Ottoman Empire was founded by Osman I (1258–1326 CE) in northwestern Anatolia in 1299. The Ottomans may have encountered gunpowder weapons either during their invasions in the Balkan in the late fourteenth century or from their Islamic neighbors.[65] By the time of the fourth Ottoman sultan, Bayezid I (1360–1403 CE), the Ottomans began using firearms in warfare. They used cannons in numerous sieges from 1394 to 1440. The Ottomans were on par with the Europeans in artillery development.[66] The Ottomans were particularly keen to capture Constantinople, the Byzantine capital, and attempted to do so several times between 1394 and 1402, and then again in 1422. But their cannons were not successful against the strong walls of Constantinople.

By the time Mehmed II assumed power, the Byzantine Empire was left with only Constantinople as its major city.[67] Mehmed II's most important remaining objective was to conquer Constantinople. Keen on artillery and military science[68] and understanding the importance of openness to new technologies, Mehmed's plan to conquer Constantinople became possible when a Hungarian cannon maker defected from the Byzantine side. He helped Mehmed's forces develop large cannons, including one giant one that was about nine meters long and could fire projectiles that weighed up to 800 kilograms. Large guns fired for almost two months, after which the walls of Constantinople finally fell.[69] Ironically, the Byzantine emperor had declined the services of the Hungarian because it seemed too costly.

After defeating the Byzantine Empire in 1453, Mehmed continued his conquests in other parts of Europe. He took particular care to strengthen his army's artillery power. He established a cannon foundry in Constantinople. He conquered Serbia, Morea, the Black Sea coast, Wallachia, Bosnia and Albania. After Mehmed, the Ottomans ended the Mamluk sultanate in Syria and Egypt and the Hungarian kingdom. The Ottoman Empire became one of the most powerful states in the world.[70] In all these conquests, firearms played an important—if not decisive—role.

Later European Developments (from 1450 to 1700)

The evolution of guns followed divergent paths in Europe and China during the period between 1450 and 1700. Guns grew larger and longer in Europe, but not in China. However, size was not the only dimension of divergence in gunpowder technology. Several European innovations changed the course of the development of guns. A few important ones are the classic cannon gun, breech-loaded guns and corning.

Classic Cannon Gun

The classic cannon gun developed in Europe around the 1480s. Longer, lighter, more efficient and more accurate than previous guns, the classic cannon gun had a new shape that endured for the next two and a half centuries.[71] Prior guns had a length eight times the width of the muzzle. The same ratio of the new guns was forty to one. This greater length gave the gun more time to convert gunpowder energy into acceleration and consequently distance and accuracy. The new gun also used iron instead of other materials. For example, an iron cannonball is four times denser than a marble cannonball of the same diameter. Because kinetic energy is half the mass times the velocity squared, the iron cannon ball packed ten to thirty times more kinetic energy than a stone cannon ball, depending on the density of the latter. The new guns were also lighter, allowing greater efficiency in manufacturing, transportation and usage.

Breech-Loaded Guns

Breech-loaded guns are those in which the projectile is loaded into a chamber at the rear side of the gun barrel, as opposed to muzzle-loaded guns in which the projectile is loaded through the front of the barrel. Breech-loaded guns had two main advantages compared to early firearms, which were mostly muzzle-loaded guns. First, it reduced loading time, enabling gunners to fire more frequent shots. Second, the crew who reloaded the guns were not exposed to enemy fire because they could reload from the rear side of the gun. Breech-loaded guns were developed in Europe in the late fourteenth century.[72] This important innovation was used extensively until the eighteenth century.

Corning of Gunpowder

Corning is a technology that produces gunpowder in granular form, unlike original gunpowder, which was very fine. Corning increased the burning rate of gunpowder, and thus made it more powerful than fine gunpowder.[73] Corned powder was more stable during storage than its predecessor because it absorbed less moisture, an important attribute in most European environments.[74] In addition, due to superior ballistic properties of corned gunpowder, even smaller guns using corned powder could render substantial destruction. This discovery probably led to the adoption of smaller and easily mobile guns in Europe from the sixteenth century onward.[75] It helped in the transportation of armies across Europe, and from Europe to the Americas and Asia, and is likely to have facilitated European colonization. The technique of corning was so effective that it was prevalent until the late nineteenth century.

Diffusion to Persia and South Asia

In addition to the Ottoman Empire, two other empires rose probably due to their superiority in firearms—the Safavid dynasty in Persia (1501–1736 CE) and the Mughal empire in India (1526–1857 CE). Some scholars term these three empires—the Ottoman, the Safavid and the Mughal—gunpowder empires, due to the significance of gunpowder weapons in the establishment and growth of these empires.

Safavids

Shah Ismail (1487–1524 CE) established the Safavid empire in 1501. Initially, the Safavids did not use firearms because they felt it was cowardly and dishonorable to use them.[76] However, in 1514, Shah Ismail, the founder of the Safavid empire, faced off against the Ottomans, who won with the help of artillery in the battle of Chaldiran. This defeat made Ismail realize the importance of firearms, and he immediately set out to establish a division of musketeers.[77] Ismail reached out to Venice using diplomatic relations to obtain firearms.[78] By 1517, Ismail's army had 8,000 musketeers, and by 1522, the number increased to about 20,000.[79]

Ismail's successors also used firearms. Ismail's son Tahmasp I (1514–1576 CE) defeated the Uzbeks in the battle of Jam in 1528, in which artillery played an important role in the Safavid victory.[80] Tahmasp's son Mohammad Shah (1532–1596 CE) obtained guns from Russia in the 1580s. But the Safavids were still reluctant to fully integrate artillery into the military. Mohammad Shah's son Abbas I (1571–1629 CE), perhaps the strongest ruler of the Safavid dynasty, reformed the army in the late sixteenth century. He made artillery prominent by integrating the musketeer corps and the artillery corps to the Safavid army. By the early 1600s, gunners made up almost 50 percent of Abbas's army.[81] Having reformed the army, Abbas took on the Ottomans in 1603 and recaptured the erstwhile Safavid capital, Tabriz, from them. He then attacked the Ottoman forts in Erivan, Armenia, with his improved artilleries and was able to conquer it in 1604.[82] Under Abbas's leadership, the Safavid were able to regain most of the territories that they had lost to the Ottomans over the last century or so.[83]

Mughals

How did the small invading force of Mughals succeed in invading India and building an empire against domestic forces that were superior in numbers and deeply entrenched? Gunpowder technology, mostly imported from nations to the west of India, played a critical role. Babur (1483–1530 CE), a descendant of Genghis Khan,[84] established the Mughal empire (1526–1857 CE) in north India. He entered India with an array of firearms that included cannons, swivel guns, mortars and matchlock muskets against the incumbent ruler Ibrahim Lodi's forces in Panipat. Even though Lodi's army was numerically superior to that of Babur, Lodi did not possess any artillery. Babur's firearms were critical to his victory over Lodi.[85] Babur had two Ottoman master gunners, Mustafa Rumi and Ali-Quli, at his disposal. Ali-Quli helped Babur deploy the Ottoman tactic of wagon fortress, which involved a network of chained carts loaded with guns and cannons providing cover for musketeers to freely fire at enemy troops.[86] Lodi's forces succumbed to artillery fire, and Ibrahim Lodi died in the battle. Babur thus clinched a decisive victory that cleared his way to conquer Delhi. In 1527, he again used firearms successfully against the Rajput forces in the battle of Khanwa.[87]

Babur took special care of his gunpowder weapons and personally motivated his master gunners.[88] After the emphatic victories at Panipat and Khanwa, Babur's opponents dreaded attacking the Mughals directly. After Babur's death in 1530, his son Humayun became the second Mughal emperor. But Humayun faced strong opposition from Sher Shah Suri, who briefly replaced the Mughal empire in north India. Sher Shah Suri defeated Humayun in 1540, and Humayun fled to Persia. While in Persia, Humayun convinced Safavid ruler Shah Tahmasp to support his campaign to recapture Delhi. In 1555, Humayun conquered Delhi to restore the Mughal empire once again.

Once Humayun's son Akbar ascended the Mughal throne, Akbar focused on developing a formidable military with a strong cavalry unit and powerful artillery. Akbar mostly procured artillery from the Ottomans, Portuguese and Italians.[89] His firearms were far superior to those the Indian defenders possessed.[90] With better firearms, Akbar won many victories and expanded and firmly established the Mughal empire in India.[91]

However, the Mughals probably did not directly innovate in or produce all their gunpowder and weapons. Instead, they purchased many of them from nations to the west of India. Even then, they focused primarily on matchlock weapons made of brass and bronze. The Mughals fell behind Western nations in technology in the late seventeenth and early eighteenth centuries when the technology shifted to flintlock muskets and cast iron artillery.[92] Thus, the Mughals were unable to withstand the Europeans who invaded India with these superior weapons. Small numbers of Portuguese, French and especially English were able to get a beachhead in South India, overcome larger armies than their own and ultimately rule the subcontinent with the help of superior weapons.

Drivers of Innovation

The above history of gunpowder helps answer the important questions posed at the outset of this chapter about the development and diffusion of this innovation: the dominant driver of this innovation that seems apparent in the above narrative is competition. But empowerment and openness also played critical roles.

Competition among Nations

Intense military competition among rivals stimulated innovation and technological advances in gunpowder. Such competition existed in China from the start of the development of gunpowder until about 1425. At about this time, the Ming dynasty was well established in China and initiated a period of relative peace. In contrast, warfare in Europe intensified with the advent of gunpowder, with constant battles and wars, as summarized above. At least one of these wars lasted for more than a century.

For example, the Song dynasty (960–1279) in China was frequently challenged by its neighbors in Asia, triggering innovations in gunpowder (see Figure 4.1). These neighbors, who also fought against one another, were themselves effective military powers.[93] The constant intense competition among these political rivals stimulated innovation, especially in gunpowder weapons. The gunpowder innovations of the Song were in no way isolated from the innovations of its foes: the Liao, Jin, Xi Xian and Mongols. Communication about rivals' innovations in gunpowder and the witnessing of battle successes due to gunpowder led rivals to quickly adopting any new innovation in the technology.[94] When a competing region adopted this innovation, it developed it further. Given these warring states were relatively similar in resources, a state of dynamic competitive equilibrium emerged, in which the rival states battled each other, capitalizing on every innovation in gunpowder technology. Gunpowder innovation was also at the heart of the defeat of the Yuan (the Mongol dynasty in China) and the rise of the Ming dynasty, established by the Yongle emperor, Zhu Di. This competitive environment was a strong stimulator of innovation. Thus, competitive rivalry among regions and states in China between 900 and 1300 stimulated constant innovations in gunpowder.[95]

After Zhu Di vanquished most of his enemies and firmly established the Ming dynasty over most of China, China did not face major wars within its territory or major threats from outside the empire. Thus, China began a period of relative calm. The only major war was the suppression of the Mongol onslaught in 1449. Correspondingly, during this time, China did not make much progress in gunpowder technology.[96] In contrast, during this period, west Asia and Europe suffered intense inter-nation rivalry, as described above. One study estimates that there were 29 wars involving England, 34 wars involving France, 36 wars involving Spain and 25 wars involving the Holy Roman Empire.[97] As a result, during this period, Europe witnessed great progress in gunpowder technology, including the development of the three major innovations discussed above: the classic gun, breech loaded gun and corning. China fell behind.

Figure 4.3 shows the number of wars in China and Europe between 1340 and 1910. Several important conclusions can be drawn from this graph. First, until about 1450, the number of wars in China were above the number in Europe, though there is constant variation from year to year. During this time, China made considerable progress in gunpowder weapons. Second, from about 1450 to about 1600, the number of conflicts was much higher in Europe than in China, except for one year. This period witnessed the first great divergence when Europe went ahead of China in gunpowder innovations. The classic cannon gun, breech-loaded guns and corning described above were all developed in Europe but not in China. Third, from about 1644 to 1839, the level of conflict was much higher in Europe than in China. This was the period when China was ruled by the Qing dynasty and witnessed a second period of relative peace. This was the second great divergence when Europe made many greater innovations in gunpowder weapons than China. During this time, less competitive warfare occurred in Chinam resulting in less innovation in gunpowder and guns relative to the prior three centuries. Thus, political competition, war and conflicts among nations seemed to be a major stimulant to innovations in gunpowder technology. These wars and conflicts reflected the level of competition among nations.

Ultimately, some European powers grew strong enough to challenge the Chinese in the Opium Wars. The Chinese efforts to suppress the opium trade, in which British merchants illegally imported opium mainly from India to China, was the

Figure 4.3 Intensity of wars in Europe and China between 1300 and 1900

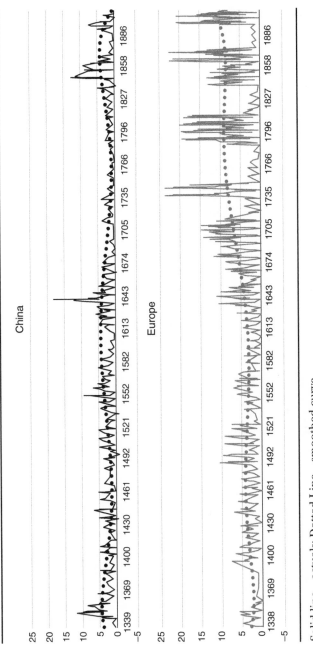

Solid line – actuals; Dotted Line – smoothed curve.

Source: Data from Graph A 2.1 of Frank Tallett, *War and Society in Early Modern Europe*, 1997.

main motivation for the Opium Wars. In the first Opium War (1839–42), Britain fought the Qing dynasty in China, and in the second Opium War (1856–60), Britain and France together attacked them. In both wars, the Chinese had to agree to unequal treaties and conceded considerable commercial and territorial privileges to the British and French due to the then superiority of the gunpowder weapons of the Europeans. In these wars, the Chinese, the inventors of gunpowder, had to face the power of their own original innovation, now advanced by the Europeans. These wars significantly weakened the Qing dynasty, and it gave way to republican China in the early twentieth century.

Empowering Engineers and Innovators

Rulers (or ruling elites) played a critical role by empowering various subjects to adopt and innovate in gunpowder technology. If the rulers embraced the new technology and promoted it within the state, the technology flourished and the nation gained a big competitive advantage against rivals that did not. The embrace was not merely verbal but also had to be concrete, matched with the assignment of financial and other resources to the technology. However, if the rulers refused to support it—or worse still, made a concerted effort to suppress it—the technology was not adopted and the nation suffered, either immediately or in due time. The example of Orban the Hungarian illustrates the role of empowerment in the development and adoption of gunpowder weapons.

Orban the Hungarian

Orban, or Urban, was a Hungarian, who famously played a critical role in the fall of Constantinople. He was a cannon expert, who had likely learned the technology of developing large guns from the Germans. During the Ottoman siege of Constantinople in 1453, Orban offered his services to the Byzantine emperor, Constantine XI. But Constantine declined his services, either because he could not pay for them, did not appreciate the value of large guns or lacked the materials for building them.

Unhappy with the Byzantine emperor's decision, Orban offered his services to Mehmed, the rival Ottoman ruler. Mehmed hired Orban for a generous salary and gave him all the resources he needed to build large guns, including one of the largest of that time.[98] In three months, Urban built a gun that was about nine meters long and could fire stone balls that weighed between 550 and 800 kilograms.[99] Apart from this "monster" gun called the *Mahometa*, he also helped develop 50 large guns and 500 smaller guns.

The guns fired for almost two months, after which the walls of Constantinople finally fell.[100] The Byzantines also had guns, though they were much smaller than those of the Ottomans.[101] Urban's giant gun may not have been the decisive factor in the Ottoman victory because it had to be cooled for about three hours after every shot and could fire only a few times a day.[102] However, on the whole, the Ottoman's gunfire augmented by Orban was far stronger than that of the Byzantines and played a critical role in the fall of Constantinople.

Openness to Disruption from Gunpowder

Throughout the history of gunpowder, no single nation developed all or even most of the innovations. Most nations that developed gunpowder did so after first encountering the innovation in an enemy or a neighboring state. That nation then adopted the innovation and in many cases, advanced it. Thus, openness played a critical role in the adoption and advancement of gunpowder technology. Such openness was particularly important because gunpowder disrupted the social and military order, rendering old military techniques obsolete and requiring the development of new skills and techniques. Failure to be open to new ideas, tools and methods in this technology meant immediate or ultimate military and political failure. A testament to this principle is the histories of the Song, Jin, Mongols and Ming in China; the warring states of Europe; and the Ottomans in Byzantine, discussed above. The importance of openness was accentuated because of the clear advantage from even small innovations in gunpowder technology coupled with the low barriers to adopting them. Two examples of the samurai in Japan and the Mamluks in Egypt dramatically show the failure of ruling groups who opposed gunpowder.

Japanese Samurai

The samurai were Japanese elite warriors who rose to power around the twelfth century, excelling at sword warfare. A small army of 10,000 samurai famously repulsed an invasion of 40,000 Mongols in 1274. In 1543, two Portuguese adventurers arrived in Japan with primitive guns. The Japanese were so impressed that they embraced the technology and developed it assiduously, having world-class guns for a while.

However, during the seventeenth and eighteenth centuries, the samurai turned against guns. The reason for their turnaround was the fear of loss of the skills with the sword that they had acquired over generations, which had also become the basis of their power and prestige in society. In contrast to the sword, which required skill, courage and discipline, the deadly gun demanded much less than the sword from the person who wielded it. Any minimally trained peasant could challenge the power of the samurai with a gun. Therefore, the samurai-controlled government restricted gun production, then required a license for owning, issued licenses only for government employees and finally reduced orders until Japan was without guns. Reluctant to empower lower classes and non-skilled warriors, the samurai ruling elite blocked the adoption and further development of gunpowder innovation.

However, around the nineteenth century, powerful Western gunpowder empires again threatened Japan. This threat left the Japanese no option but to adopt gunpowder to preserve the country's independence.[103]

Mamluks in Egypt

The Mamluks originated during the reign of the Ayyubid sultanate (1171–1260). The Ayyubid dynasty enslaved young boys from central Asia and brought them up to be fearless warriors. These slave-soldiers soon became the knighted nobility of Muslim society throughout the Ayyubid dynasty and eventually took over the sultanate in Egypt. They ruled there from 1260 to 1517. The Mamluks were a powerful ruling elite of knights, who fought with a sword while mounted on a horse. Like the samurai, the Mamluks opposed the adoption of gunpowder because fighting with the gun required little skill and minimal training, unlike fighting with a sword on a horse. So, the Mamluks considered the gun an undignified and inferior weapon and—similar to the samurai and other ruling classes—blocked its adoption to preserve their status and political power.[104]

For the Mamluks, political power rested on their strict military training and upbringing that idolized the sword. Additionally, strong social ties manifested in a brotherhood in arms. The long training resulted in special battlefield skills and military capabilities. However, using gunpowder technology was comparatively simple and did not require as many years of training or as complex capabilities. An infantry soldier using a gun required much shorter training than a swordsman jousting on a horse. The adoption of gunpowder technology in Mamluk Egypt would have meant a shift in political power away from traditional Mamluk elites toward lower classes—and the Mamluks had no interest in empowering lower classes.

In contrast to the Mamluks, the Ottoman sultan empowered lower classes over the warrior elites in the Ottoman Empire. In particular, he empowered the Janissary, a gun-bearing infantry, to adopt and master gunpowder technology. One of the results was the Ottoman victory over the Mamluks in a series of battles between 1516 and 1517 in which the Janissary infantry defeated the Mamluk cavalry.

Thus, more than most other technologies, gunpowder was one that diffused relatively easily and could be adopted or rejected by various rulers and peoples. Those that adopted, adapted, and innovated it, flourished. Those that rejected it, failed. The forces of openness, empowerment, and competition facilitated the adoption, adaptation, and innovation of gunpowder.

5

Golden Age of Chinese Water Navigation

The period from 1403 to 1433 marks China's golden age in water navigation. Under the patronage of the fourth Ming emperor, Zhu Di, China assembled a massive naval force, larger and technologically more sophisticated than any the world had ever seen then or until at least a century later. The ships embodied numerous innovations in navigation and gunpowder weapons well ahead of any other country in the world then. That technology enabled the fleet to make seven voyages to lands around China, Ceylon, India, Saudi Arabia and the west coast of Africa. China may have had the capacity to disover the New World about a half-century before Portugal and Spain.

Yet by 1434, all this was history. The ships were grounded, shipyards shut down, records of the expeditions destroyed, maps probably burned and the golden age of Chinese navigation came to an abrupt end. By the time of the seventh and last great voyage in 1433, China was on the verge of discovering and mapping the whole world, 60 years ahead of the Europeans. Then, suddenly, China ceased to be a world power. This chapter describes some of these innovations and attempts to explain the rise and fall of Chinese navigational power.

Innovations of the Chinese Golden Age

Between 1403 and 1433, the Chinese fleet may have required the construction of over 2,000 ships.[1] The fleet's ships were bigger and

Figure 5.1   Model of Zheng He's treasure ship

Source: From Hong Kong Science Museum, eWilding/Shutterstock.

longer than any built before in China. Fleets of about 270 ships with crews of about 27,000 consisted of five categories: treasure ships carrying gifts and dignitaries; horse ships carrying horses; supply ships carrying grain, water and other supplies; billet ships for troop transport; and combat ships.[2] Large ships may have had up to nine masts and 12 sails and been up to 440 feet long and 180 feet wide (see Figure 5.1).[3] The largest of these ships dwarfed those of the Portuguese explorer Vasco da Gama, who was the first European to navigate from Europe to Asia some 60 years later (see Chapter 7). Still, the Chinese ships had a displacement more than 10 times that of da Gama's ships.[4] Indeed, one study concludes that "the largest [fifteenth-century Chinese] ships may well have been the largest wooden ships ever constructed."[5]

The fleet's ships were the most advanced in the world at the time and combined several Chinese innovations such as gunpowder weapons, the compass, movable sails, a retractable balanced rudder, multiple watertight compartments, and dry docks for ship construction.[6] Some of these innovations, like multiple watertight compartments, retractable balanced rudder and dry docks, were adopted in Europe only centuries later.[7] Gunpowder weapons included brass and iron cannons, mortars, flaming arrows and exploding shells[8] (see Chapter 4). Another innovation, adopted from the Arabs, was the astrolabe to measure the distance between sea level and the North Star to determine latitude.

A major innovation was the combination of two types of earlier ships: one for sailing in shallow waters and the other for sailing in choppy oceans.[9] The new ships adopted elements from both types of ships in order to increase their navigability, stability and endurance in the open seas. These ships could sail in turbulent ocean waters and avoid hitting a shoal. They constituted both a form of transportation and a fighting force. However, the fleets' sheer superiority in size, prowess and numbers did not necessitate any major battles with enemies. The fleets mostly served for diplomacy and show, demonstrating China's maritime power.

Admiral Zheng He (1371–1433) was a eunuch of Muslim origin,[10] handpicked by Emperor Zhu Di to construct and lead the fleet. From 1405 to 1433, he personally led or launched seven voyages (see Figure 5.2). These voyages reached Champa across the South China Sea, Vietnam, Java, Sumatra, the Straits of Malacca, Ceylon (now Sri Lanka), the Malabar Coast (Cochin and Calicut in West India), Persia, the Arabian Peninsula and the East Coast of Africa. Some authors speculate that the voyages may have reached the northern coast of Australia and parts of the Americas.[11]

These voyages became the embodiment of Chinese technological superiority in oceanic navigation. The voyages themselves spurred further maritime innovations. Zheng He and his crew documented the landscape and important features of the coasts they visited. This documentation led to the creation of new valuable maps, navigational charts and astronomical data that may have diffused to and influenced later navigators in Venice and Portugal (see Chapters 6 and 7).[12]

Drivers of China's Rise in Water Navigation

Chinese navigational innovations peaked during the reign of the Yongle emperor Zhu Di (1403–24). Understanding the

Figure 5.2 Voyages of Zheng He

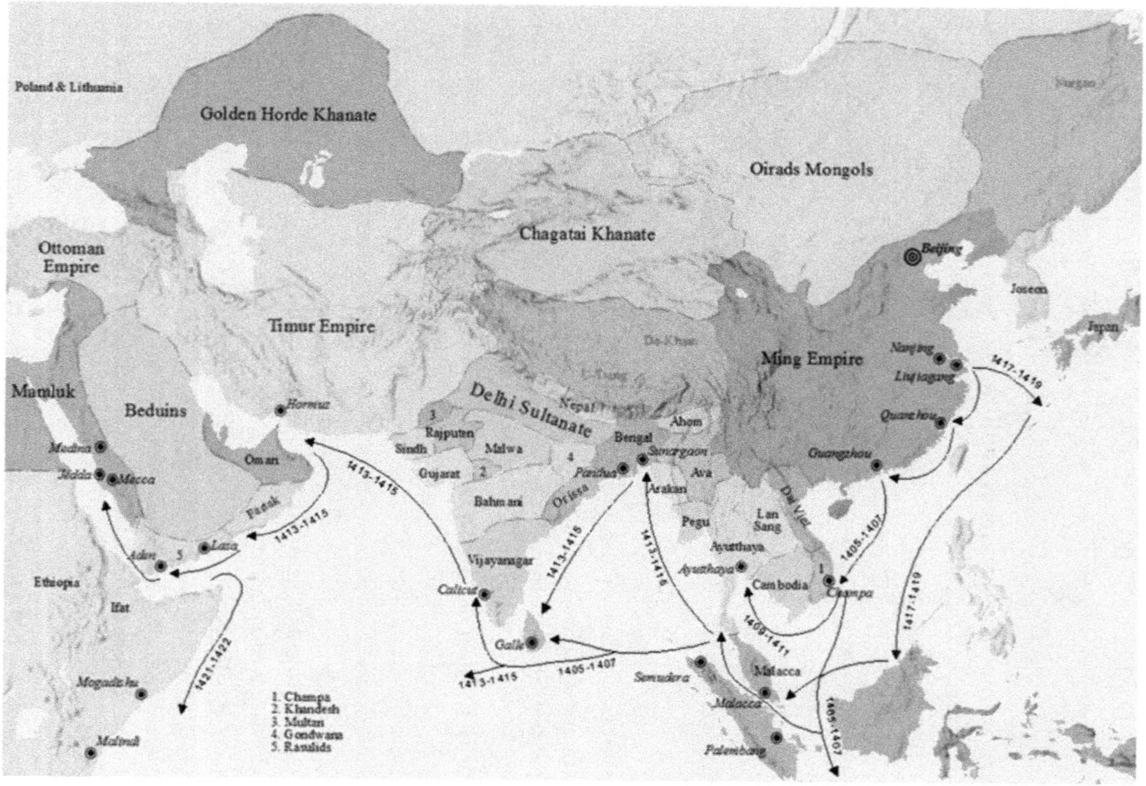

Source: Seasonsinthesun/Wikimedia Commons.

circumstances of his rise to the throne sheds light on the factors that led to this flowering of innovation and its subsequent decline.

Emperor Zhu Di was the fourth son of the Hongwu emperor, who founded the Ming dynasty by overthrowing the ruling Mongol Yuan dynasty. The Hongwu emperor ruled from his capital in Nanjing in the south of China. He appointed Zhu Di, prince of Yan, to his capital in Beiping (modern Beijing) in the north of China. Zhu Di's charter was to expel the last remnants of the Mongols and prevent their reentry. He initially accepted the appointment of his eldest brother and later his brother's son, the Jianwen emperor, as succeeding emperors in Nanjing. Zhu Di himself may have been overlooked by his father, the Hongwu emperor, because his Mongol mother was supposedly pregnant with him before she married his father.[13] As a result, Zhu Di suffered from a feeling of illegitimacy throughout his life.[14]

Both the Hongwu emperor and his grandson, the Jianwen emperor, aligned with the mandarins, who were scholar-bureaucrats steeped in Confucian philosophy. The mandarins believed in tradition, order, continuity and stability. They were the dominant political class in China outside of the imperial palace. Within the palace, eunuchs held sway, notorious for their intrigue and politicking. As with most bureaucrats, the mandarins were suspicious of innovation because it invariably led to change, if not disruption and transformation.[15] Another reason for their commitment to order and continuity was that innovation and change contradicted their strong believe in Confucian philosophy.

When the Jianwen emperor ascended the throne, he sought to exterminate his uncles, such as Zhu Di, who were potential rivals to the throne. Zhu Di resisted this move and a three-year civil war ensued between the forces of Emperor Jianwen (ruling from Nanjing) and those of Zhu Di (ruling in Beiping). After some initial setbacks, Zhu Di's forces ultimately reached the outskirts of Nanjing. There, with the help of the eunuchs, he entered the city, crushed the forces of the Jianwen emperor and was anointed emperor. The prevailing Jianwen emperor fled into exile, never to reappear again.

Openness to Internal and External Innovations

Confucianism is a Chinese ethical, sociopolitical and philosophical system developed from the teachings of the Chinese philosopher Confucius (551–478 BCE).[16] Confucianism permeated every aspect of Chinese society early on. As a result, Chinese society was typically quite hierarchical and rigid in structure.[17] Confucianism's focus on order and tradition imposed considerable obstacles to innovation, especially when Chinese society adopted Confucianism as its moral code. According to Confucianism, foreign travel interfered with family duties. Likewise, trade was inherently debasing and exploitative,[18] so innovation and business ownership were not considered respectable occupations. Thus, becoming a successful merchant, builder or innovator did not mean advancement in social status for mandarins. Thus, throughout most of China's history, Chinese society frowned upon such trades.

Despite these strong social and cultural patterns, some Chinese rulers had an ardent desire for foreign goods and displayed them at court. Foreign tradesmen—mostly Arabs and Persians—brought these goods to China from across the Indian Ocean and South China Sea.[19] Some rulers were fascinated by exotic goods foreign merchants brought to China. Therefore, despite Confucianism's discouragement of trade and tradesmen, some Chinese emperors were open to sailing, tradesmen and foreigners. Thus, when Confucianism was weak, innovation flourished, but when Confucianism was prominent, innovation stagnated.[20]

Emperor Zhu Di was one such ruler. He was suspicious of and had a lifelong distrust of the mandarins, who opposed his rise to power and supported his chief rival. As a result, he had little commitment to the mandarins' Confucian philosophy. Instead, he remained open to new ideas and classes of people, especially the eunuchs, who helped his rise to the throne. As one author remarks, "Religious tolerance was one of Zhu Di's great virtues, and the junks [ships] also habitually carried Islamic, Hindu and Buddhist savants to provide advice and guidance."[21] Under the emperor's patronage, Zheng He established a language school in Nanjing to train interpreters who traveled with the fleet, enabling communication with people of multiple languages including Arabic, Persian, Swahili, Malayalam and Tamil.[22]

Emperor Zhu Di's personal philosophy led him to embrace and advance innovations, irrespective of their origin. These included adopting innovations already known in China or neighboring kingdoms, and creating innovations by combining and enhancing these. He had few qualms about maritime trade and travel. During his reign, one of the most important expressions of Chinese openness was Chinese disposition toward the sea. He strongly favored maritime trade and both diplomatic and economic interactions with foreign powers and cultures. He chartered the building of the navigational fleets. He encouraged the six great naval expeditions that occurred during his reign under the admiralty of Zheng He. Thus, during the reign of Emperor Zhu Di, innovations flourished, especially in water navigation. These innovations were probably the "culmination of a long tradition of Chinese seafaring in the South China Sea and Indian ocean."[23] Due to this openness to innovations, the Chinese navy greatly expanded, and China entered its golden age in water navigation.

Competing with Enemies, Real and Imagined

The manner of Emperor Zhu Di's ascent to the throne had a major imprint on his reign. He suffered from a perception of illegitimacy due to his suspected birth and rebellious rise to the throne. He was plagued by a fear of real or imagined competitors, who threatened his empire or challenged his place on the throne.[24] The Mongols in the north were the major enemy. However, chief among Zhu Di's rivals to the throne was his nephew (and previous Jianwen emperor), whom he had driven into exile and obscurity. After Zhu Di's rise, rumors persisted of the survival and location of the Jianwen emperor—in monasteries, mountains or in lands across the oceans. Zhen He's biography states that Emperor Zhu Di himself feared that Emperor Jianwen "had fled beyond the sea."[25] Thus, driven by real or imagined competition, throughout his reign, Zhu Di sought to build grandiose enterprises to establish his legitimacy and the superiority of his reign to competing nations around China. These enterprises included the development of a massive navigation fleet in the south, the building of the Forbidden City (a massive palace in present-day Beijing), and the rebuilding and expansion of the Great Wall of China and the Grand Canal. Part of the early motivation for the naval expeditions may have been to find and destroy Emperor Jianwen.[26]

These enterprises employed millions of workers, artisans, engineers and other talent throughout the empire. Three of these enterprises were likely a net drain on the treasury: construction of the Great Wall, navigational fleets and the Forbidden City. However, the rebuilding of the Grand Canal and the expansion of the water network in central China led to a significant gain in agricultural productivity and transportation efficiency. Moreover, the overall building effort probably led to a huge stimulus to the economy with productivity gains in innovation, employment, transportation, trade, defense and construction. For example, at the opening of the grand Forbidden City, Emperor Zhu Di ordered one of Zheng He's expeditions to bring to Beiping dignitaries from Vietnam, Malaysia, Indonesia, Ceylon, Cochin, Calicut and elsewhere. These dignitaries were transported, hosted and dispatched in grand style and showered with many valuable gifts. As a result, they were in awe of the technological progress, wealth and might of the emperor and his empire.[27]

Thus, real or imagined competition was an effective driving force in the technological advances during the reign of Emperor Zhu Di.

Empowering Eunuchs

Traditionally, the eunuchs had authority within the palace but not in the nation's administration, which the mandarins controlled. In a major departure from the past, Emperor Zhu Di entrusted the construction of these enterprises to the eunuchs, who helped his rise to the throne.[28] He ignored the authority of the mandarins, who supported the previous emperor. Zhu Di's empowerment of the eunuchs to build and govern his enterprises is a classic case of innovations flourishing when sponsored by outsiders, rather than by the incumbent ruling insiders, the mandarins.[29] Zhu Di was not encumbered by tradition and Confucianism, as were the mandarins. Rather, he was driven to fulfill his own grandiose enterprises.

The epitome of Zhu Di's philosophy in terms of empowerment was his appointment of the Muslim eunuch Zheng He to build and lead the massive navigational effort. The goal of the latter was to impress on nations of the contemporary world the rise, power and wealth of Emperor Zhu Di. More than anything else, the giant fleets and the seven expeditions were a demonstration of Chinese power. The fleets established Chinese presence overseas, dominated and kept open maritime trade, and impressed on foreign rulers the superiority of Chinese technology and the legitimacy of the rule of Emperor Zhu Di.[30]

Thus, Emperor Zhu Di's empowerment of the outsider eunuchs at the cost of the incumbent mandarins turned out to be a shrewd strategy in bringing change, creating transformative enterprises, advancing innovation and achieving his goals. However, it may not have been a sustainable strategy. The mandarins were a powerful and entrenched class who soon reasserted their authority once Emperor Zhu Di died.

China's Decline in Water Navigation

After the death of emperor Zhu Di (1424), the Confucian mandarins regained power at the imperial court. Under their influence, his son continued the suspension of the naval voyages and the building of additional ships. Zhu Di's grandson allowed a seventh expedition in 1433, led by Admiral Zheng He. However, after the early death of Zhu Di's grandson, subsequent emperors—under the influence of the mandarins—put a complete halt to all navigational construction and expeditions. With that, China's golden age of navigation ended abruptly. The seventh naval expedition in 1433 turned out to be the last. China ended its immense maritime progress at its peak.[31] The ships were grounded, shipyards shut down, records of the expeditions destroyed and maps probably burned.[32] The age of innovation that had yielded fascinating naval advances and taken China to diplomatic heights and oceanic dominance gave way to stagnation and complete seclusion from the outside world. What were the reasons for this dramatic change in national policy? Several factors probably played a role.

First and most importantly was the demise of Emperor Zhu Di and his philosophy, which contradicted that of the Confucian mandarins. The flourishing of navigational innovations arose primarily from the philosophy of Emperor Zhu Di toward openness, empowerment and global competition. However, this was a top-down philosophy, imposed by the emperor during his reign. As in other periods throughout Chinese history, it did not permeate all the institutions of the empire, most of which were run by the mandarins. When Emperor Zhu Di died in 1423, the power briefly passed on to his son and then his grandson. After the death of the latter, the new heirs to the throne were not so beholden to the eunuchs nor driven by the same passions of Zhu Di. Rather, they fell under the sway of the mandarins. The mandarins, who had suffered a loss of power and prestige to the eunuch navigators, convinced the new emperor to end the naval expeditions and shut down the entire naval enterprise. The conservatism, traditionalism and inward focus of the mandarins—steeped in Confucianism—triumphed over the openness, empowerment and global extravagance of Emperor Zhu Di.

Second, external competition was a factor. From about 1433, China entered a period of relative peace, when its main enemies were mostly defeated (see Figure 4.3). The seven expeditions established China as the superior naval power, all the way from Japan and Korea in the east to Java and Sumatra in the south, to India, Persia and Africa in the west. China faced no serious competition from the seas. The only competition it faced was from Mongol forces in the north. In particular, until the Portuguese arrived on the shores of China some 90 years later around 1521, China faced no major threats from the ocean. This lack of competitive threats from the seas decreased the need for advances and innovation in navigation. The supremacy of Chinese naval technology lulled the subsequent emperors and mandarins into a false perception of permanent security. Had the competitive threat of the Portuguese occurred 90 years earlier, the attitude of Chinese emperors and mandarins toward the naval fleets may have differed.

Third, costs played a role. Emperor Zhu Di did not use the fleet to colonize and exploit foreign nations, as did the Europeans a century later. One major reason for this behavior was that Chinese philosophy perceived China as the center of the world and an occupation of "barbarian" peoples was not necessarily a desired goal. Rather, the fleets were often involved in generous diplomacy, lavishing gifts on foreign rulers and transporting their ambassadors to and from Beijing in style and luxury. The goal was to demonstrate power and gain political and cultural influence. The goal of the expeditions was primarily to awe nations into submitting to Chinese political authority.[33]

Trade flourished during the seven expeditions. The main exports of the treasure fleet were silk, porcelain, tea and jade products, though they also carried iron, iron products, hemp, wine, oil and candles.[34] The treasure fleets brought back to China then exotic goods including silver, spices, sandalwood, precious stones, ivory, ebony, camphor, tin, deer hides, coral, kingfisher feathers, tortoise shells, gums, resin, rhinoceros horn, sappanwood, safflower, cotton cloth and ambergris.[35] At times, the import of some of these goods (like black pepper) was so high, that they went from precious goods to commodities. However, many of the exotic imports may have reached only the wealthy and not the masses, who paid the costs of the naval buildup through taxes. Nevertheless, the mandarins counted all these exotic imports as extravagant luxuries and not necessities.

At the same time, the cost of constructing, launching and managing the seven expeditions was massive. Building the fleets involved enormous costs in terms of natural resources such as wood and iron, labor and land for construction.[36] The design of the massive treasure ships evolved from river barges used for transportation in the rivers and canals of the interior. Thus, they developed to be large, broad-beamed, square-shaped, flat-bottomed ships.[37] While suitable for carrying large armies of soldiers and transporting dignitaries in luxury, they may not have been as efficient as the smaller Portuguese caravels that came a century later for transporting goods (see Chapter 7).

Estimates of the costs of only constructing—not running—the large treasure fleet range from 5 to 15 million piculs of grain (a picul would equal about 130 pounds).[38] Peasants had to pay taxes for these massive costs, either indirectly through grain or directly through supplies for the construction of the ships. Moreover, many of the eunuchs involved in the administration of taxes were corrupt, siphoning off goods or funds for their personal gain rather than the welfare of the state.[39] One author estimates that 300 acres of teak forests were required to build one large treasure ship.[40] Construction of the whole fleet may have required felling millions of acres of forests, resulting in a revolt in Vietnam, from where some of the wood was obtained.

Thus, the costs of building the navigational fleets and running their seven epic expeditions were huge, while their benefits rarely trickled to the taxpayer. Still, to the mandarins the costs loomed large, while the gains were discounted. Steeped in conservatisms and tradition, they failed to see the potential of trade and were fearful of the power of the eunuchs, who built and managed the fleets. So, after 1433 when new emperors came to power, the mandarins killed the entire navigational enterprise.

Fourth, internal competition was likely a factor contributing to the end of China's golden age. During this time, China was a relatively monolithic political entity. A single emperor ruled over a vast empire. His decision was law over that whole empire. The main internal competition was between the eunuchs and the mandarins. When the mandarins overpowered the eunuchs, a single ruling against navigation enterprises became unquestioned dogma throughout the entire empire. There was little opportunity for dissent, experimentation and innovation. To appreciate this monolithic political entity, one can compare the situation in China versus that in Europe. Here, geography had created numerous city-states, kingdoms and national states in constant rivalry and competition. If one hesitated or stalled, innovation leadership was picked up by another. Indeed, just in navigational innovations between 1400 and 1700, leadership passed from Venice to Portugal to the Netherlands to England (see Chapters 6, 7, 8, 9 and 10). Internal competition within Europe stimulated innovation. Lack of internal competition within China led to stagnation and isolation.

Thus, while China shut down its naval expeditions and closed in on itself, precisely the opposite forces were at work in Venice and Portugal. Indeed, these two nations took off where China stagnated (see Chapters 6 and 7).

6

Venetian Shipbuilding: Mastering the Mediterranean

In 1204, under the disguise of the Fourth Crusade, a Venetian force sacked Constantinople, the capital of the great Byzantine Empire, greatly weakening it.[1] Despite their oath not to attack fellow Christians, crusaders under Venetian command looted, terrorized and vandalized the city for three days. They destroyed or stole many ancient Roman, Greek and Byzantine treasures. Among the stolen works were the famous horses of the Hippodrome, which were shipped back to Venice and today adorn St. Mark's basilica there. Crusaders also destroyed the great library of Constantinople. Additionally, they raped and massacred thousands of the city's inhabitants. Ironically, the Fourth Crusade originally sailed from Venice to protect Christendom (and Constantinople) from the Muslims. In effect, the assault on Constantinople greatly weakened the Byzantine Empire, led to its decline and probably contributed to its fall to the Ottomans two centuries later[2] (see Chapter 4).

Ironically, Venice was the western outpost of the Byzantine Empire, which gained its independence from the empire only in 814. An agglomeration of marshy islands, Venice lacked the landmass for agricultural produce, which was the primary source of wealth in those times. How did this relatively poor backwater grow to become the capital of an empire that ultimately brought Constantinople to its knees?

This chapter argues that innovations in shipbuilding and finance were primarily responsible for the rise of the Venetian empire from the eleventh to the fifteenth century. Venice

Figure 6.1   Venetian galley

Source: Marzolino/Shutterstock.com.

developed the technology for designing and efficiently building the Venetian galley, the most important ship in the merchant and military navies of the state (see Figure 6.1). The galley was built in the Arsenal, which at its peak could turn out about a galley a day (see Figure 6.2). The Arsenal was most responsible for Venice's rise as an empire, distinguishing the state from all other city-states in the Mediterranean. Merchant galleys carried its trade, while military galleys protected it against enemies. Financial innovations enabled risk sharing and triggered merchant entrepreneurship in trade. Trade was the other major source of Venice's wealth and power.

Between 1400 and 1450, at the peak of its power, Venice had about 36,000 sailors operating 3,300 ships. Venice handled trade between Europe and Asia in many goods including spices, silk, jewelry and cotton. At its peak, the Venetian empire (Stato da Mar) included parts of northern Italy (Domini di Terraferma) plus several colonies, including Istria, Dalmatia, Albania, Negroponte,

Figure 6.2 Venetian Arsenal

Source: Julia Panchyzna/Shutterstock.com.

Morea, the Aegean islands of the Duchy of the Archipelago, Crete and Cyprus.[3] The Venetian empire established several key ports throughout the Adriatic Sea, including Alexandria, Tunisia, Thessalonica, Tripoli, Tyre, Athens and Constantinople. Their ships even traveled west to Barcelona and Valencia, and north to London and Bruges.

In addition to these ports, the Venetians had trade treaties with several foreign powers that added to their wealth. For example, Venice's economic relationship with Egypt, Syria and states on the eastern Mediterranean shore were tied to the Mamluks—the Islamic rulers centered in Cairo. The Mamluks were the middlemen between South and Southeast Asia and Venice, trading goods such as textiles, spices, precious stones and paper.[4] Mamluk metalwork, with inlaid silver and gold, was exported to Venice and decorated homes both in the city-state and other parts of Europe. Later, the Venetians had trade relations with the Ottoman Empire. Although Venice competed with the Ottomans over territory, it generally sought peace for the sake of trade. The

Ottomans exchanged raw silk, cotton and ash for products from Venice's glass, paper and textiles industries.

This chapter describes first the innovations and then the drivers that contributed to the rise of Venice from a marshy set of islands to the dominant power in the Mediterranean and in global trade by the fifteenth century.

Venetian Innovations

Venice's most important innovations were shipbuilding and various financial innovations.

Great Galleys and the Arsenal

Venetians were traders and, like the rest of the Mediterranean, utilized a ship known as the galley to transport merchandise to and from the various ports of the empire. These large ships, distinguished by their long, slender hull, had sails but were primarily propelled by rowing. This allowed the galleys to move swiftly and precisely by depending on manpower, not the unpredictable winds or currents.

There were two main ways in which Venetian shipbuilding differed from its competition and predecessors. First, the Venetians revolutionized shipbuilding by creating separate galleys for trade and war, and constructing each differently from the past. Second, they established a government-run shipyard, known as the Arsenal, where the construction process was very advanced for its time and resembled a modern assembly line. Both these innovations are explained below.

The Venetians produced two types of galleys—the "great galley" (*galie grosse*), designed specifically for carrying goods on trading voyages,[5] and the light galley (*galie sotili*) for war.[6] The great galleys were invented between 1294 and 1298, and came into common commercial use within a decade.[7] Both ships were initially very similar but diverged over time in proportions and rigging. The great galleys were developed to store goods and protect them from pirates. Various innovations enabled these ships to become massive, enabling them to carry a crew of more than 200 and weighing as much as 250 tons. The light galleys, somewhat similar to the Vikings' longships, were tailored for war. The length of their hulls was eight times the width, and the deck was less than six feet above the keel. These proportions greatly facilitated speed and maneuverability.

Venetian galleys differed in construction from those of Venice's Viking and Byzantine competitors. The latter constructed the hulls of their ships by first binding heavier, overlapping boards to make a watertight hull and then supporting this planking by adding ribs and braces.[8] A major Venetian innovation was to first construct the skeleton of the ship (the keel and the ribs) and then nail the planks to the frame.[9]

Research suggests that the development of this "rib and plank" construction required several innovations. This method of building had previously been used only for river-bound vessels. Adapting this method of construction for use in the open waters of the Adriatic and Mediterranean Seas required a lot of experimentation.[10]

The construction of the skeleton had several benefits. First, this form of construction dramatically reduced the cost of shipbuilding. It required less skilled carpentry and ultimately used less wood.[11] Second, the ships could be easily repaired and cracks easily filled. A mixture of fiber and pitch allowed for the creation of a watertight hull—one that was cheaper to produce but just as seaworthy as its predecessors. The abundance and strength of these galleys enabled the Venetians to fend off pirate attacks and ensure that goods arrived at their designated ports.

The other major Venetian innovation was the development of the Arsenale Nuovo around 1320 from the older Arsenale Vecchio. While previously used to maintain ships, the Arsenale Nuovo was built and allowed for the construction in one place of both the states' navy and merchant ships. This state-run shipyard was the base for the mass production of Venetian galleys. The workers in the Arsenal were highly skilled and specialized in various areas of shipbuilding, enabling great efficiency. Work was organized into three stages of production, each developing its own guild: the shipwrights, the caulkers and the oar makers.[12] The shipwrights (*marangoni*) produced the "live work"—the keel, ribbing and frame of the ship.[13] The caulkers (*calafati*) fastened the planks to the frame, built other super structures and filled in (caulked) any open seams. The oar makers (*remeri*) produced the oars for the galley. These specialists worked with smaller specialties to arm the galleys and attach the rigging. This standardization and division of labor allowed for the rapid production of ships. At its peak, workers in the Arsenal could efficiently construct ships at the rate of one galley in less than a day.[14] At any given time, the Arsenal aimed to keep a reserve of 100 light galleys and 12 great galleys.[15] In addition to ships, the Arsenal also produced weapons of war, including in later years, gunpowder weapons. The speed and organization of construction in the Arsenal surpassed that of any competing nation or state, making Venice the foremost naval power in the Mediterranean and in global trade until it was displaced by the Portuguese (see Chapter 7).

Between the thirteenth and fifteenth centuries, Venice's merchant and military navies enabled it to dominate trade in the Mediterranean, colonize Mediterranean ports and islands, and bring immense wealth and prestige to the city-state. It also enabled the state to attack and briefly occupy Constantinople, the seat of the Byzantine Empire, of which Venice was at one time merely the westernmost colony.

Financial Innovations

Venice's primary source of income was its trade between Europe and the Byzantine Empire, whose capital was Constantinople. Through Constantinople, Venice had access to the goods of the East, including silk, cotton, jewelry and spices. An important aspect of this trade was the risk involved in shipping in those times. The round trip from Venice to Constantinople would take about three months.[16] If successful, such trips incurred considerable profits. At the same time, such trips also incurred considerable risks.[17] First, they required a large amount of capital to finance the ship, crewmen and goods. Second, along the way, voyages could incur pirates, travel hazards and delays, which could mean partial or total loss for the merchants involved in trade. Third, such trade involved buying goods on credit, with no guarantee of returns. The collateral, in the form of goods, left port with the ship. As a result, an agency problem emerged because the residing merchants had to trust the traveling merchants, who could abscond with the goods for a one-time gain.

To resolve these problems, Venice established an innovative contract, called the *colleganza*, which was a first in the Mediterranean countries and Europe.[18] Features of this system became a precursor to the subsequent limited liability company that characterized business development in the Netherlands, Britain and other parts of the world (see Chapter 8). The contract specified two agents: the residing merchant and the traveling merchant. The residing merchant provided the capital for the goods and received a share of the profits. If the contract was unilateral, the residing merchant received 75 percent of the profits; if it was bilateral, the residing merchant received 67 percent of the profits.[19] The traveling merchant undertook the risks of the journey and the responsibility for its completion, earning the remaining share of profits. The residing merchant's risk was limited to the value of the goods. Losses beyond the goods were borne by the traveling merchant.

The *colleganza* split the risk of trade between a residing merchant and a traveling merchant. The residing merchant was typically someone wealthy enough to finance the trip and run the risk of losing all his investment for at least a huge profit. In contrast, the traveling merchant was someone who, though poor, was willing to undertake the risky journal and ensure its safe completion. If the trip was successful, the reward was substantial without any investment of capital. Successful traveling merchants, after accumulating sufficient wealth, could then rise to the class of residing merchants.

The technological and financial innovations allowed Venice to become the major trading center in Europe and the Mediterranean from 1000 to 1500. Without the shipping innovations, Venice would not have had the merchant ships to carry the trade, and the military ships to defend against pirates and competitors. Without the financial innovations, Venice would not have had the sharing of risks and the incentivization of merchant entrepreneurship. Thus, these innovations enabled Venice to amass great wealth, becoming, by the fifteenth century, possibly the richest city in Europe.[20]

Drivers of Innovation

Three institutional drivers played a key role in the advances made in Venetian shipbuilding and financial innovation: competition, openness and empowerment.

Empowering Innovators and Merchants

Several critical events between 1000 and 1200 triggered the empowerment of the merchant class in medieval Venice.[21]

In the early ninth century, Venice was the western corner of the Byzantine Empire. In 814, as a reward for help in countering the Carolingian empire, the Byzantine Empire granted Venice independence, with a doge as ruler. For the first few centuries after that, the doge was a monarch of unlimited powers. Selection was partly hereditary, as the doges came from just three families.[22] However, in 1032, due to prior political unrest, a wealthy merchant was elected doge, bringing an end to the system of hereditary doges from a few privileged families. Moreover, their powers were increasingly limited. As trade became a driving force in the economy of Venice, the power of merchants rose with respect to that of the doge. This in turn led to laws that encouraged trade and the amassing of even greater wealth and power for the merchant class.[23]

In 1082, in return for the Venetian military support in fighting the Normans, Constantinople signed a treaty, called the Golden Bull, with Venice. Among other concessions, this treaty let the Venetians (1) trade throughout the Byzantine Empire without taxes; (2) control the main harbor facilities of Constantinople, with control of several public offices; and (3) define a trading district within Byzantium, for shops, a church and a bakery.[24] These concessions greatly enhanced Venice's trading advantage relative to its competitors in Italy and the Mediterranean. It triggered the steady increase in wealth and power of the traders in the Venetian Republic.

In 1172, Doge Vital II Michele headed an armada to battle Constantinople and press for the release of hostages. That expedition not only failed but also was blamed for bringing back the plague. On his return, the Doge was assassinated and a new constitution was written, empowering an elected parliament called the Great Council. In the next couple of decades, three major changes made Venice more egalitarian. First, the doge had to be elected by members of the Great Council. Second, upon his election to office, the doge had to take an oath not to expropriate state property or to preside over cases against himself. Third, in all major matters of state, the doge had to consult with a six-member Ducal Council, which itself was elected by the Great Council.

These institutional changes empowered innovators and fostered numerous innovations. In particular, the two centuries after 1082 witnessed the enactment of numerous financial innovations including the *colleganza*, primary markets for debt, secondary markets for debt, equity and mortgage instruments, bankruptcy laws, double-entry accounting, deposit banking and a reliable medium of exchange (the Venetian ducat).[25] These innovations, in turn, further encouraged trade, brought new entrants into the merchant class (thereby facilitating the empowerment of individuals through social mobilization) and generated great wealth for individuals and the state.

Most importantly, the merchant class was, initially at least, not restricted to the rich and the privileged. It was open to those willing to take the risk of becoming a traveling merchant. As a result, the *colleganza*, which democratized risk and the opportunity to become wealthy, further increased the empowerment of entrepreneurial individuals. Due to it, "the riskiness of trade together with the widespread involvement of Venetians in this trade, created a great deal of income churning—mostly rags to riches but also some riches to rags."[26]

Intense Competition with Trading Rivals

Intense external competition with trading rivals fueled Venetian innovations, especially in shipbuilding. Venice depended greatly on trade and in turn on the power of its navy. It competed fiercely in trade with its principal Italian rivals Genoa, Pisa and Amalfi.[27] Venice's naval prowess caused it to win the favor of Constantinople. After defending Constantinople from an attack by several powerful Carolingians, Venice received the favorable trade treaty—the

Golden Bull—in 1082 (see prior section). The treaty allowed the Venetians to establish a Venetian colony in Constantinople itself (the first foreign state afforded this privilege), effectively blocking its rivals from trading within the city.[28] In effect, the treaty gave Venice a monopoly on trade within Constantinople, shutting out its rivals Genoa, Pisa and Amalfi. The relationship between the two states ended, however, in 1171 when the Venetians' desire to monopolize trade started a riot—a protest of the extension of trading rights to Pisa and Genoa. The Venetians were expelled from Constantinople. In January 1183, the emperor of Constantinople allowed the Greeks to massacre the remaining Italians in the city (namely Genoese and Pisan citizens) as a way of repaying his supporters. In doing so, the emperor started a war with the Pisans and the Genoese. With Constantinople's other enemies, the Normans, arming themselves, the emperor had allowed the Venetians to reenter the empire and take over the Venetian quarter in exchange for protection. However, upon return, their special privileges were revoked.[29]

The reentry of the Venetians into Constantinople was followed shortly by a call by Pope Innocent III to launch a crusade into Jerusalem. The pope and his allies in France and Spain approached Venice to request a vessel to transport soldiers into Jerusalem. The pope sent envoys to request Venetian assistance for two main reasons. First, Genoa and Pisa, the other two states with naval capabilities, were at war with one another and unable to help a crusade.[30] Furthermore, Venice had assisted the papacy on a crusade in 1121, and Pope Innocent hoped that the newly appointed doge, Dandolo, would be equally willing to help. The doge was the elected ruler of the Venetian republic. Under the fiscally shrewd doge's guidance, Venice agreed to join the crusade for monetary purposes, aiming to earn an enormous amount of money for their other projects. In addition to promised payments from the crusaders, the Venetians hoped to wipe out middlemen traders in Cairo and Alexandria, creating a monopoly of trade in the eastern Mediterranean and eliminating competition with Pisa and Genoa.

Thus, innovation in Venice's shipbuilding industry—specifically that which increased the agility and war capabilities of its ships—was motivated by a desire to beat out its trade competition. Venice's warships included flying bridges that could be launched up out of the ship as a means for breeching walls. These bridges were later employed against Constantinople as a means of recovering lost investment when the crusaders could not pay for the ships they ordered. The crusaders and the Venetian victory over Constantinople allowed the Venetians to dominate trade in the center of the former Byzantine Empire, and the Venetians' strong galleys ensured that their wares were protected.[31]

Through the years, Venice's biggest competitor was Genoa, a fellow Italian city-state. Just as Venice was somewhat isolated from the Italian mainland by its lagoons, so Genoa was separated from the Italian mainland by a wall of mountains. The rivalry between the states dated back to at least the thirteenth century. Though Genoa was slower to expand than Venice, it was equally invested in trade in the Mediterranean. Four open military conflicts occurred between the two states. The last of these Genoese wars, also known as the War of Chioggia, was a victory for Venice. Both states' naval fleets and economic resources were seriously depleted, as demonstrated by Venice's generous concessions in the Peace of Turin—between Genoa and Venice in 1381—that brought the war to a close.[32] However, because of strategy, efficient shipyard production and the prioritizing of certain trade routes, Venice was able to recuperate and reconstruct its fleet. The reconstruction enabled Venice's ultimate success over the Genoese fleet and its ability to be the preeminent trading power in the Mediterranean.

Another of Venice's almost constant competitors was the Ottoman Empire, as early as 1300.[33] The Ottoman Turks had previously challenged Constantinople, and, when Venice took over several of Constantinople's ports in 1328, the Italian city-state inherited this rivalry. The Ottoman Empire continuously expanded over the Balkans, and, for a period, it looked as though all Aegean ports were going to be occupied by the Turks.[34] The Ottomans had yet to develop a navy, and most of their attacks on these port cities were overland. Many Greek states, fearing Ottoman attacks and acknowledging their inability to defend themselves, willingly offered seaports for sale in exchange for protection by Venice. By increasing the sizes of their fleets, purchasing sea ports and making treaties with the purpose of delaying armed conflict, the Venetians were able to expand their control of the Aegean Sea.[35]

This strategy worked until about 1470, when the Ottomans constructed a fleet and attacked Venice's main base in the Aegean Negroponte. In 1479, the Venetians surrendered to the Ottomans, turning over Negroponte and a couple other Aegean islands. This battle marked a turning point in Venetian maritime history—while they were still an active presence, they no longer dominated the seas and no longer "held the gorgeous East in fee."[36] Especially after the Ottomans formed an alliance with the Barbary pirates, Venice was unable to maintain a navy that could overcome the Turks.

Openness to Diverse Peoples and Traders

From early times, Venice was a relatively open city, while religious dogmatism, persecution and parochialism were common across contemporaneous Europe. The Venetian city-state was founded as a safe haven for people escaping persecution in mainland Europe after the fall of the Roman Empire.[37] During the period of its rise, Venice was a fairly open city-state. Because it was a state that depended on trade, Venice was fairly welcoming to foreign merchants, financiers and workers. The Venetian government made provisions for housing and protection for merchants operating in its domains. Except during wartime, the state even allowed the ships of its competitors, including Genoa, to enter port at San Nicolo,[38] in Venice. Due to Christian teaching, money lending was considered unchristian, and some Jewish financiers were allowed to operate in the city.[39] The Republic taxed the Jews and, in 1516, restricted their living quarters, thereby creating the first ghetto. However, the fact that German, Spanish and Levantine Jews came to Venice, in the fourteenth and fifteenth centuries, indicates that Venice's level of tolerance of Jews was perhaps higher than in other European cities.[40]

Venice's openness was also due, in part, to its need to replenish its population following outbreaks of the bubonic plague. The first epidemic was in 1348. In each of these epidemics, large portions of Venice's population were wiped out—each of these epidemics killing approximately one-third of the population.[41] Because of the dramatic drop in population and the need to maintain a workforce, the Venetian government encouraged immigration to revive business. Most immigrants came from the Italian mainland, but sailors and seamen from Greece and Dalmatia also migrated there. These workers became thoroughly Venetian, and the skilled workers were rewarded with citizenship. Immigrants could obtain citizenship in two stages, subject to certain conditions.[42] After 10 years they were entitled to partial citizenship, and after 25 years they could get full citizenship. Noncitizens had rights according to the guild they belonged to and their rank in it.[43]

The story of Zaccaria Stagnario is an example of the openness to foreigners and the empowerment of people in Venice.[44] His grandfather was a Croatian slave who was freed when his Venetian owner died. His father was a helmsman. In 1199, he traveled in a *colleganza* to Constantinople, which yielded a big payoff. By 1206, he held office as a counselor to the first Venetian podesta in Constantinople and was rich enough to be a residing merchant in two *colleganzas*. He was welcomed into the ruling elite, and his descendants served in the Great Council in nearly every session from 1261 to 1295.

Another person highlighting the openness of the Venetian spirit was Marco Polo (1254–1324), a Venetian merchant and traveler. He is famous for his trip to China to explore the source of the goods that Venice traded in. Accounts of his 24-year travels sparked Venetian and European interest.

Venice developed an information network with an inspection system that controlled movement in plague-infested areas, preventing the plague from spreading to other parts of the city. So doing, made the city-state a haven for people in other parts of plague-infested Europe.[45] The benefits of citizenship, the compensation and rewards of becoming a craftsman, and the escape from the bubonic plague in other parts of Europe all led to an influx of immigrants and talent. Venice's shipbuilding during most of the 1400s produced a number of important Greek and Dalmatian shipwrights and designers.[46] However, this situation changed dramatically by the seventeenth century when Venice's government barred foreign shipwrights and limited foreign visitors, as Venetian citizens were "more inclined toward love of their fatherland."[47] Such a change may have also contributed to Venice's decline (see later subsection).

Meritocracy and the Empowerment of Individuals in the Arsenal

Because of the importance of the Arsenal, the Venetian Senate regulated the recruitment, training, work and promotion of individuals to ensure innovative and high-quality output. Initially, the artisans enrolled in the Arsenal were strictly limited, with only a few enjoying the right to work in the shipyards when they pleased. Later, this policy was made more liberal. The regulation of employment in the Arsenal established a meritocracy for work and promotion. The state appointed a group of lords and commissioners (*patroni e provveditori all' Arsenale*) to run the Arsenal and supervise the munitions, maintenance and construction branches.[48] The lords and commissioners were frequently changing, only in office for one to four years and were frequently men of naval experience. These lords and commissioners appointed foremen—one for each of the main areas of production, carpenters, caulkers, mast makers and oar makers. The foremen were all masters of the craft that they supervised.[49] Each year, the master craftsmen were reviewed to evaluate their performance and determine wages and promotions.[50] These four foremen worked alongside approximately 30 other "gang bosses" to manage production. The gang bosses were frequently selected based solely on training and shipbuilding skills. When one position became available, all interested masters were invited to partake in a competition to demonstrate their skills in a formal examination.[51]

While these competitive examinations took place, they were infrequent for foremen positions, namely because shipbuilders were rarely considered for foremen's posts until their expertise was firmly established in the Arsenal.[52] Applicants for top offices chosen to partake in foreman examinations were frequently responsible for innovations or ideas that expedited the shipbuilding process.[53] Candidates for high-level foremen positions were also distinguished if they had experience beyond the Arsenal's walls. For example, before completing his exam, Zorzi di Christofolo had spent almost no time in the Arsenal. Chosen to become the foreman of the oar makers, he had spent over 20 years working with the navy.[54] Foremen positions were also open to younger individuals, again emphasizing the importance of expertise and the ability to take an active hand in construction even at the foreman level. An example of this was Zuane Luganegher becoming foreman shipwright at only 24.[55] The reasoning behind this was that younger foremen were in their "creative prime," so hiring them prevented technical skills and ideas from being out of date.[56]

Decline

The most critical driver of the decline of Venice was the Portuguese caravel, which obsoleted the Venetian galley (see Chapter 7). In 1498, the Portuguese, led by da Gama, sailed around the Cape of Good Hope at the tip of Africa to reach India. He did so with the help of the caravel, an innovative Portuguese ship that relied on sails rather than oars for propulsion. In contrast, the Venetian galley depended primarily on oars rather than sails for propulsion. This change in innovation had a dramatic impact on the fortunes of Venice. Previously, Europe imported goods from China and India through Venice. The main imports from Asia were spices, cotton, silks and jewelry, which were then exported to Europe. The imports from Europe were woolen products, metals, salt, books, wine, wheat and millet, which were then exported to the Muslim world and Asia. Venice's own production and exports included ships, arms, silk products, mosaics, draperies and glass.[57]

Venice relied on numerous middlemen over land and sea routes in the Byzantine Empire, Syria, Turkey, Egypt (Mamluks), Arabia and the Mediterranean and Arabian Seas. All these middlemen added time, costs and uncertainty to the trade. Uncertainty also arose due to the vagaries of weather and storms, the wars, the pirates and thieves, and the change in regimes. Da Gama's sea route to India eliminated all these middlemen. Thus, it reduced costs, shortened time and increased reliability in trade. Venetians understood that they had acquired their wealth and power by trade between Europe and Asia. All of this was now jeopardized by Portugal's sea route to India and the East.[58]

Moreover, this Portuguese discovery rendered the primacy of Venice obsolete. In fact, not only did Venice suffer from the cutting of middlemen, but also the entire cross-Mediterranean trade network suffered greatly as soon as the Portuguese activated the new trade route. In 1511, the Portuguese conquered Malacca on the Malay Peninsula, also known as the Spice Islands. Tome

Pires, a Portuguese adventurer, commented, "Whoever is lord of Malacca, has his hand on the throat of Venice."[59] This dramatic change was brought about by the Portuguese innovation of the caravel, an ocean-going ship that obsoleted the Venetian galley (see Chapter 7).

Venice's failure to foresee and develop the sea route to the east may be partly because it had over invested in the Arsenal and the galley and partly because the forces that led to its rise, openness, competition and empowerment, were in slow decline in the fourteenth century. With Venice's rise, wealthy merchants acted to gradually change the laws in the republic to restrict membership into the ruling class to an increasingly small group of elites, who accounted for less than 10 percent of the population. These changes in laws limited entry of new merchants, entrepreneurship and competition.[60] Moreover, over time, new Venetian regulations from 1376, 1476 and 1483 further limited and eventually forbade any non-Venetian from engaging in commerce with the Ottoman Empire, including Ottoman citizens residing in Venice.[61] Earlier in its history, entry and competition were the primary drivers of the fertile climate of innovation and economic progress that marked the rise of Venice. New restrictions hindered free entry and competition and with the decline in these drivers, innovation suffered. The wealth of the merchant class blinded them to the drivers of that wealth. In a myopic turn, the merchants shut down the drivers of their wealth, competition and the empowerment and acceptance of newcomers. As a result, subsequent innovation and progress suffered. Realizing its rapid decline and the increasing dominance of Portugal, Venice reinstated partial privileges to foreign merchants in 1524. Realizing that these merchants were an important source of income to the state, further actions to reinstitute the previously forged privileges continued until the 1600s.[62]

However, these changes were too little and too late. Portugal and not Venice developed the caravel to exploit the sea route to the east. Portugal innovated where Venice failed, reaping the enormous fruits of that innovation (see Chapter 7).

7

Portuguese Caravel: Building an Oceanic Empire

On May 20, 1498, Portuguese sailor Vasco da Gama and his crew landed at Calicut, India. They were the first Europeans ever to set foot in India after an ocean trip. They arrived safely after an unbelievably long voyage across the Atlantic and Indian oceans lasting eleven and a half months. After almost reaching the coast of Brazil, da Gama navigated his way back to Africa, landing 130 miles from his intended destination, the Cape of Good Hope. Through this Cape, he finally reach Calicut.

This journey epitomized the cutting-edge techniques in water navigation developed by the Portuguese in the fifteenth century. Da Gama had been able to choose a route that avoided coastal currents to take advantage of the westerly gales of the south Atlantic as they made their way back toward Africa. Da Gama achieved this feat by expertly executing navigational techniques with the instruments of the period.[1] Portugal's maritime innovations and discoveries led to it becoming one of the first multicontinental empires during the Age of Discoveries from 1400 to 1600. Portugal became the first country in Western Europe to trade directly with India, Malay and the West Indies by sailing vessels. The subsequent Portuguese empire covered Brazil and parts of Africa and South, Southeast and East Asia—as far as Nagasaki and East Timor.[2] Portugal charted more of Africa and the Americas than any of its contemporaries.

Along with the increased trade in West Africa, the Portuguese established colonies in Madeira in 1419 and Azores in 1427, two islands far into the Atlantic off the Iberian coast.[3] The Portuguese instituted sugar plantations on Madeira and Azores. The sugar trade proved to be immensely lucrative in Europe, financially sustaining their exploration of the Atlantic and other potential oceanic trade routes.[4] They were the first of the European states to enslave African natives whom they encountered on their voyages, exploiting enormous profit both from the trade of the slaves and from the plantations on which the slaves worked. Portugal was also the first country to establish a sea route around the Cape of Africa to Malacca, Goa and Bombay. In doing so, the Portuguese were able to monopolize the thriving Eastern spice trade, replacing the Venetians, who had depended on a complicated network of overland routes and middlemen to procure spices in the Mediterranean.[5] As evidence of the wealth attained through such trade, the Portuguese, at their peak, were bringing in about 4–5 million Portuguese cruzados (between 8.4 and 10.5 tons of gold) annually from their various colonies.[6]

Christopher Columbus, who sailed under the Spanish flag, had been heavily influenced by his experience in Portugal, where he learned much about navigation.[7] In the early fifteenth century, Portuguese trade in the Atlantic consisted of slaves and gold from West Africa and sugar from the island colonies of Madeira and Azores. These activities laid the foundation for the major innovations in navigation and subsequent discoveries of routes to the New World and to Asia by Columbus and da Gama respectively.

What innovations enabled the small nation of Portugal to foster this first successful sea journey to India and then build a global empire? What were the drivers of those innovations? Why did Portuguese leadership in innovation and world trade decline? This chapter addresses these issues.

Portuguese Innovations

The key Portuguese innovation was the caravel in addition to other innovations in shipbuilding, navigation and cartography (see Figure 7.1). The early champion of these innovations was Prince Henry the Navigator, the third child of King John I.

Figure 7.1 Portuguese caravel

Source: Michael Rosskothen/Shutterstock.com.

The Caravel

In the 1420s, when Prince Henry championed a quest for a sea route to India, he focused on perfecting the design of Portuguese ships. To actualize his mission, Prince Henry established the school of navigation at Sagres, on the southwestern tip of Portugal (see Figure 7.2). The school focused on improving not only the ships but also navigation, cartography, astronomy and oceanography.[8] Prince Henry championed navigational innovations, taking a personal interest in their design.

The principle navigational innovation of the Portuguese was a ship called the caravel—a vessel for trade along the Atlantic coast, designed in the early fifteenth century. Borrowing from the technological advances of the Arabs in the 1400s, the caravel had three masts, lateen sails and a sternpost rudder.[9] Lateen sails were triangular sails attached at 45 degrees to the mast instead of traditional square sails. The ideas of multiple masts and the

Figure 7.2   Lighthouse of Cabo Sao Vicente, Sagres, Portugal

Source: Andrew Buckin/Shutterstock.com.

sternpost rudder were adapted from the Arabs, who may have learned these techniques from the Chinese. Arabs and others used triangular "lateen"-style sails in the Mediterranean Sea and the Indian Ocean. The Portuguese themselves developed a system of using groups of small sails for each mast that were more manageable than earlier designs, allowing the crew to trim the sails to size for various changes in winds, sea conditions and weather. In addition to these features, the caravel also had a forecastle (an area at the front of the ship but below the deck, normally used for the crew) and a deckhouse (a structure built on top of the deck for storage).[10]

Gavin Menzies writes of Prince Henry's contributions to the design of the caravel, "But of all the improvements Henry introduced, none surpasses his brilliant adaptation of the Arabic lateen sail. The later caravels had lateen sails on mizzen and main masts and a square sail on the foremost, [which] could be converted at sea to be either square- or lateen-rigged. A mariner could sail southwards from Portugal, square-rigged before the prevailing winds, then convert to lateen sails to return north into the wind. Although tiny in comparison to Chinese junks, the caravels were much more nimble and maneuverable."[11]

The caravel designs went through various innovations over the years. The first design was the Caravela Latina with a leaner hull and three masts that were all lateen rigged. This innovation was followed by the Caravela Redonda, a four-mast vessel—one mast with a square sail accompanied by three masts with lateen sails that improved the ship's maneuverability. In addition to these innovations in sails, the caravel was lightweight with a shallow draft that helped ships navigate swiftly while exploiting the strong winds of the Atlantic (and later Indian) Ocean. It also had a large cargo to hold supplies for long voyages. Many of the ideas for the ship were brought to the Iberian Peninsula through the large number of Arabs living there as well as the Peninsula's proximity to the Middle East. Accordingly, the Spaniards borrowed from and used the Portuguese caravels as well. Two of the ships that Columbus took to the new world—the Niña and the Pinta—were caravels.

Guns and Gunpowder

Another major innovation adopted by the Portuguese was gunpowder for military use. After gunpowder reached Europe (see Chapter 4), the Portuguese embraced it in the fifteenth and sixteenth centuries to gain naval and commercial advantage.

The major innovation of the Portuguese in gunpowder technology was equipping the caravel with guns. Initially the guns were placed on the upper deck, but that proved to be highly unstable for the ship. Later the guns were placed on the lower deck. The most important of these placement innovations was ports at sea level, through which the guns could fire. Not only did this placement make the ship more stable, but it also hit the enemy ships at their sea level, where they were most vulnerable to flooding and capsizing.

Indian and Arab merchants had been plying the Indian Ocean for thousands of years before the Portuguese arrived. But they had not developed a vessel equipped with guns and the maneuverability of the caravel in the high seas. The Arab ships in the Eastern waters were generally galleys, which conducted naval combat by ramming into enemy ships and trying to board them by grappling hooks. This method of fighting proved futile against the Portuguese ships, which carried cannons and fought with gunfire from afar.[12] Once the Portuguese reached the Arabian Sea with the caravel equipped with guns, they were able to dominate the Sea and set up trading ports along the west coast of India. Because of their superior ships and firepower, the Portuguese were able to prevent any traders from conducting business in the Arabian Sea or the west coast of India without their explicit permission.[13]

Other Major Innovations

As with the caravel, the Portuguese were opportunistic with the technology of other nations. They borrowed and built upon these foreign designs, mastering and modifying them for use at sea. One example of this is the development of the mariner's astrolabe. The astrolabe—an Arab innovation—had been in use for many years to determine latitude and plot location based on the position of astronomical objects like the sun. The Portuguese modified it in the late 1400s, adjusting its size and weight to counter the ship's movement.[14] Similarly, the quadrant, an instrument for determining the latitude of astronomical objects, and the "cross-staff," a device for measuring angles, were modified for nautical use in those same years. The translation of Ptolemy's *Geographia* from Greek to Latin by the Italian Jacobus Angelus in 1409 created tremendous excitement throughout Europe and was helpful in the development of correct maps. Ptolemy, recording the latitude and longitude known to him during his time, mapped Africa and India in their correct positions. Empowered by imperial support, Portuguese navigators quickly adopted these newly available findings and the method of fixing latitude in navigating. Portugal's quick adoption and adaptation of prior discoveries reflects the country's openness to foreign knowledge.

Initially, European sailors depended on portolanos, charts based on compasses, which showed coastal features and ports. These portolanos did not take the curvature of the earth into consideration. Therefore, they were not useful in navigating open oceans. However, several Portuguese innovators made valuable scientific and astronomical innovations that improved the accuracy of charts and maps. For example, the cartographer Jacob ben Abraham Cresques, a Jew who had fled from Spain to settle in Portugal, discovered the lunar cycles. Knowledge of the lunar cycles allowed pilots to navigate the strong tides south of the Cape Bojador on the west coast of Africa.[15] Dom Joao de Castro realized that iron objects affected compasses, becoming the first to record incidences of "compass variation."[16] Portuguese cartographers made maps and navigational charts more detailed and accurate than the previous portolanos as well as documented new discoveries. At the order of the king of Portugal in 1485, astrologer Mestre José created tables of solar declination that proved vital as a navigation manual for determining latitude through using the sun.[17]

Drivers of Innovation

In the fifteenth century, the Portuguese surpassed the previously powerful Chinese and Venetians in navigational expertise. The Venetians lost out due to a singular focus on perfecting the Venetian galley for navigation in the relatively tranquil Mediterranean. The Chinese lost out due to a self-inflicted seclusion by the later Ming dynasty after the death of Emperor Zhu Di (see Chapter 5). In contrast, the tiny nation of Portugal leaped forward over other Atlantic nations, such as England, Spain and the Netherlands, because of its unique openness to ideas from abroad, empowerment by visionary leaders and intense competition.

Openness to Foreigners, Jews and Muslims

During the Middle Ages, Portugal was a small, poor country. Its population was half that of England's. It lacked substantial natural resources, and poverty and famine were frequent.[18] Because Portugal was an unsophisticated agrarian society at this time,[19] the country relied on trade with other countries for economic sustenance.[20] Trade was a powerful force, as it exposed the Portuguese to outsiders and opened it to innovations from other countries. This attitude together with the need to perfect the navigational tools of trade triggered the Portuguese to innovate. This situation led it to adopt shipping designs, the compass and gunpowder from the Chinese and Islamic and Asian nations to the east.[21] In particular, three modified technologies of the Portuguese—the astrolabe, the quadrant and the cross-staff—were borrowed from nations to the east.[22]

In 1128, under Alfonso Henrique's rule, the Portuguese traded goods such as fish, salt, olive oil, wine and pine resin with northern Europe. By the fourteenth century, the Portuguese also traded jewels, pearls, spices, dates, nuts and fruits in this market. The robust trade encouraged the Portuguese to be open to other nations.

As early as 1240, Portuguese merchants began residing in France.[23] The Portuguese city Lisbon had a deep harbor and was surrounded by the Tagus Valley, geographic factors that led to its growth as a major port city. Merchants from several countries settled in the city. En route to England and Flanders, merchants from Genoa and Venetia settled in Portugal during the 1300s, resulting in the development of an Italian community. By allowing Italian merchants to partake in trade in the city and providing concessions on timber and maritime insurance, Portugal was also able to increase the size and numbers of its merchant fleets.[24] King Dinis of Portugal summoned a Genoese to be his admiral in 1317, a strategic acquisition that provided Portugal with Genoese knowledge of ships and the sea.[25]

The Portuguese also showed openness in allowing Jews and other foreigners to immigrate to the country and use their skills. Indeed, a majority of Portuguese innovations in water navigation during the fourteenth and fifteenth centuries derived from Islamic and Jewish texts. In the ninth century, the Muslim Caliph al-Ma'mūn founded the "House of Wisdom," an institution in Baghdad where Greek works such as those of Archimedes, Euclid and Ptolemy were translated into Arabic. The Muslims built on these ideas over the next several centuries and made significant scientific accomplishments—one even proposing a heliocentric solar system centuries before Copernicus.[26] These ideas may have reached Western Europe through the Islamic settlements in Spain and Portugal.

King John I and his wife Queen Philippa, who established the House of Avis in 1383, were unusually open to ideas from abroad. "Their court was one of the most enlightened in Europe, a center for men of scholarship and ability regardless of religion."[27] The tolerance and even embrace of people of non-Christian faiths was unusual at the time in Europe, where prejudice against Jews and Muslims was rampant. In contrast, Portuguese royalty hired Jewish scholars to translate Islamic and Jewish scientific texts into Latin for use by Portuguese navigators.[28] Although the Catholic Church considered such actions somewhat heretical, the benefits of religious toleration outweighed the costs.

In contrast, Spain in particular was less welcoming of Jews and Muslims. This is particularly evident in 1492, when King Ferdinand and Queen Isabella expelled the Jews from Spain. In contrast, Portugal gave the expelled Jews asylum in Portugal.[29] Until 1531, when the some Portuguese Catholics began clamoring for an inquisition, Jews were allowed to continue their practices under the label of "New Christians," after being sprinkled with holy water.[30] Many of these Jewish immigrants were merchants and scholars, who, though required to undergo conversion to Christianity to avoid persecution, provided important insights to the Portuguese on trade in Brazil, Africa and Asia. They also played an important role in scientific development in Portugal, securing trade with Muslim intermediaries as well as attracting Genoese business to Portugal.

One event in the opening of Portugal was Prince Henry's capture of Ceuta in 1415. This was an important Arab port on the north coast of Africa overlooking the Straits of Gibraltar. By taking Ceuta, Portugal learned some more from the Arabs. The Arabs revered scholarship, had preserved the classics of Greece and Rome, had traded all over the then known world and had described the sea route to India, with details of ports and cities.[31] With the conquest of Ceuta, Prince Henry realized the value of Arab learning and trade and potential further conquests. Subsequent conquests included Madeira, ports along the African coast, ports along the west coast of India and then East Asia, the Americas and Brazil.

The takeoff and fruition of Portugal's exploits and conquests brought more immigrants into the country. Italian navigators brought their skills to Portugal to exploit the lucrative slave, sugar and gold trade of the Portuguese in the Atlantic.[32] German and Italian bankers invested in Portuguese maritime activities for their own benefits. The country stayed open to these talents from other countries. All this diversity and influx created a fertile atmosphere in which the innovations of navigation and the military flourished. Over two centuries, the small state of Portugal grew into a global empire.

Empowering Innovators and Investors

Portugal of the fourteenth and fifteenth centuries tolerated, if not encouraged, a high level of entrepreneurship. Much of the Portuguese innovation in navigation was made by ordinary sailors, traders and entrepreneurs in quest of profits from the trade in gold, slaves and sugar.[33] Fishermen searched for new routes throughout the Atlantic for new sources of cod—a staple of the Iberian diet—due to the dominance of the English and Dutch in the north.[34] Cobblers used sealskin to make shoes, creating a constant demand for new seal hunting routes.[35] Demands for all these products motivated Portuguese sailors to explore the Atlantic Ocean and record their findings, leading to more accurate maps and improved navigation techniques.

Portuguese royalty empowered brilliant residents to innovate in methods of overseas navigation, sponsoring voyages to create better maps of the African coast. The exploration of the Atlantic by the Portuguese created a commercial system in the Atlantic that financially enabled Portuguese sailors to continue with their innovations and discoveries. Prince Henry chose Sagres, a promontory 200 feet above the Atlantic in southwestern Portugal, to be his base and the center of his efforts to build a navigational force for Portugal. There he built a chapel, hospital and navigational school. There and elsewhere, King Afonso V and Prince Henry empowered individuals within Portugal to conduct research and build innovations in water navigation.

For example, up to the fifteenth century, navigators created portolanos, or navigational charts, to map landmarks, coasts, ports and compass bearings for maneuvering in the Mediterranean Sea. Such charts were not nearly as accurate as those for the Atlantic Ocean because of the larger distances. In order to resolve this problem, Prince Henry hired the finest astronomers and mathematicians of Europe to develop instruments and tables by which pilots of ships could establish the latitude of landmarks along the African Coast. Documenting these locations made creating portolanos of the Atlantic possible.[36]

The expansion of trade resulting from the Portuguese discovery of a new route to Asia brought foreign investment into the Portuguese economy. German bankers dealing in silver and copper took a keen interest in the new route of the Portuguese to Asia, where silver and copper were much more valuable.[37] The German bankers, such as the Imhofs, Webers and Fuggers, all invested large sums of money in Portuguese shipping. Similarly, Italian traders indulged in the newfound slave and sugar trade, putting lots of their wealth into the Portuguese maritime navigation efforts. Foreign investment from the Italian merchants and German bankers empowered Portuguese navigators to further innovate in navigation.[38]

Competition

Some phenomena fostered an atmosphere of intense competition in Portugal, which stimulated innovation.

First, early Portuguese navigational advances were probably motivated by the need to protect against raiders and pirates from North and West Africa who attacked villages along the Portuguese coast and fled with booty and slaves.

Second, a constant and fierce competition with Muslims, pushed the Portuguese, perhaps more than any other European country on the Atlantic, to innovate in order to open and then keep control of new trade routes from Europe to Africa, Madeira, Azores and later to the Indian Ocean around the African Cape of Good Hope. For this purpose, they established various forts at points along the route and equipped these bases with gunpowder and cannons. Competition with the Muslims who mediated the gold and slave trade from West Africa and the spice and jewel trade from India motivated the Portuguese to explore the Atlantic coast of Africa in search of ways to eliminate the Muslim middlemen.

Third, prior to Portugal's exploration of the West African coast, the Arabs traded most of the gold from the region by way of camel caravans that made their way to the Mediterranean Sea.[39] These caravans facilitated trade between West Africa and the

Iberian Peninsula that brought African gold, slaves and spices to Portugal.[40] The dominance of the Arabs in the West African trade caused limited supply and created high prices of precious metals in the Iberian Peninsula.[41] The development of the lateen-sailed caravel partly resulted from Portuguese merchants' attempts to circumvent these caravans and bring goods directly from the source.[42]

Fourth, the Crusades of the Middle Ages kept a perpetual state of competition between the Catholic Europe and the Islamic East. This competition with the East led the Catholic Church to empower monarchs. For example, the first papal bull, issued in 1452 by Pope Nicholas V, authorized the king of Portugal to attack the Saracens (the Muslims to the east), enslave them and expropriate their land and wealth. Numerous other papal bulls with similar rhetoric followed.[43]

Decline

Three factors likely played a role in the decline of the Portuguese innovation enterprise: conquest, the Portuguese Inquisition and Dutch innovation.

Between 1580 and 1640, the Portuguese empire came under the rule of the Hapsburgs in Spain. The change in leadership led to a decline in the Portuguese oceanic and colonial prowess. Spain chose to neglect the Portuguese posts, defending the Spanish ones instead. Accordingly, Portugal's overseas holdings, especially in East Asia, decreased dramatically.

Another factor behind Portugal's decline was the Portuguese Inquisition. Beginning in 1500, Jews slowly began leaving Portugal due to fear of persecution. This exodus accelerated in 1531 with the call for a Portuguese Inquisition, which was a strong expression of seclusion and a lack of openness. Many Portuguese and Spanish Jews, known as Marranos, fled to the Netherlands and England. The number of Marranos in the Netherlands increased exponentially in 1609 when the Dutch government declared a policy of openness and religious tolerance. Many Marranos took their ties to the Levant and the West Indies with them, increasing and improving Dutch trade connections. The Marranos also developed the Dutch silk, tobacco, sugar and diamond trade.[44] By expanding the Dutch trade network, the Jews who fled Portugal because of the Inquisition challenged the Portuguese and Spanish monopoly and paved the way for the Dutch domination of trade. The Inquisition marked the closing of what was until then the relatively open society of Portugal. That openness facilitated the adoption and development of innovations, as explained earlier. Ironically, the Netherlands and England, the two countries that welcomed the Portuguese religious and ethnic groups fleeing the inquisition, turned out to be the countries where innovation flourished, as we shall see in subsequent chapters.

A third factor that contributed to the decline of the Portuguese empire was Dutch innovation in shipbuilding. Following the development of the caravel, with its use of lateen and square sails, less focus was directed to the improvement of sailing ability, while an increased focus was directed to the reduction of operating costs of ships—changes in capacity for cargo as well as the number of crew required.[45] The high costs of moving goods with armed ships—where the space required for soldiers and weaponry reduced the amount of space available for goods to be traded—reduced the commercial prospects for many countries, including Venice and Spain.[46]

In response to this quest to reduce operating costs, the Dutch shipbuilders introduced the *fluyt*. While not as aesthetically pleasing as the caravel or other ships with tall masts, the squat—a somewhat stubby ship with a spacious cargo hold and a system of pulleys and blocks—could make the voyage with a substantially smaller crew.[47] The *fluyt* and other armed variations of the ship met the needs of the sixteenth-century shipping industry. With these innovations, the Dutch traders replaced the Portuguese as masters of the sea (see Chapter 8).

8

The *Fluyt* and the Building of the Dutch Empire

Between 1568 and 1670, a coalition of small waterlogged and sea-threatened Dutch provinces in Northern Europe broke out to become a global empire and the dominant power in world trade. This event marked the Dutch Golden Age. How did this great shift occur so quickly in such a disadvantaged region? This chapter explains how innovations powered this transformation and investigates the factors that drove these innovations.

The shift began in 1568 when a coalition of northern provinces known as the Low Countries revolted against the Spanish rule of King Philip II, in protest of high taxes and the persecution of Protestants. Initially a mere entrepôt, or a place where commodities were stocked for reexport, the Low Countries grew into the Dutch Republic, which, by the mid-seventeenth century, was the dominant trading power in Europe. Initially a nation built on bulk shipments of grain and fish, the Republic initially expanded trade to include wool, silk and spices. As trade spread abroad, the Dutch East India Company was founded in 1602 to protect Dutch trading interests in the Indian Ocean. It steadily expanded into trade of goods from Asia and Africa, most importantly spices such as cinnamon, pepper and cloves. The East India Company built forts in colonies, maintained treaties with native rulers and became the first multinational corporation. By the end of the seventeenth century, the Dutch empire spanned from the east to west of Japan, Indonesia, Bengal (India), South Africa, New York and Suriname. By 1670, the prosperous Dutch trade totaled about

Figure 8.1 Dutch sawmill in Ulst, Friesland

Source: Jstuij/Shutterstock.com.

568,000 tons—an amount that exceeded Spanish, Portuguese, French, English, Scottish and German shipping combined.[1] The East India Company alone was worth 78 million guilders. If adjusted for inflation, this value is equal to a few trillion dollars in 2012, making it the most valuable company in history.[2]

While several important technological and financial innovations contributed to Dutch dominance in trade, innovations in ship design and shipbuilding played a crucial role in fostering its expansion. This section first describes the innovations and then the factors that promoted its development.

Dutch Technological Innovations: The *Fluyt* and the Sawmill

The key innovation that let to Dutch dominance was the *fluyt*, along with other innovations such as the sawmill, various navigational tools and the design of various business instruments and institutions (see Figure 8.1).

Figure 8.2 Replica of Batavia, example of the *fluyt*

Source: Roel Meijer/Shutterscock.com.

Initially, Dutch merchants used *karveels*, the equivalent of the Portuguese caravel. By 1550, the karveels were replaced by smaller, more maneuverable ships called *vlieboot* and *boeier*.[3] However, rapidly growing maritime trade required larger, faster and more economical ships than these. This need led to the invention of the *fluyt* in the 1590s (see Figure 8.2). Building separate ships for carrying merchandise and fighting was itself a major innovation.[4] Focusing on merchant ships enabled the Dutch to greatly reduce production, labor and operating costs, which were very high for military ships.

The design of the *fluyt* was highly innovative. Drawing from their expertise in exporting large amounts of fish and grains, the Dutch designed the *fluyt* with an extremely large cargo space, a flat bottom and a shallow draught. This design enabled the *fluyt* to navigate shallow waters of ports and rivers better than ships of other nations. Still, the ship could carry about 360 tons.

The *fluyt* was built of both pine and oak. The pine made it light, while the oak gave it strength. In addition to these features, the ship was steady in bad weather. The *fluyt* was rather long, the length being anywhere from four to six times the breadth of each ship. This length was paired with a tumbledown form—an inward slant of the part of the hull above water—as well as a pear-shaped stern, creating a low center of gravity, increased stability and improved ability to go against the wind. The ship had three masts. The foremast had a square sail, the main mast had two square sails, and the third mast had lateen sails. The ship itself was operated by a system of pulleys and blocks. This system allowed for a substantially smaller crew than the *fluyt*'s predecessors, such as the caravel. Sir Walter Raleigh, when discussing these ships, mentioned that the Dutch *fluyt* could sail with a crew of seven to eight men, whereas the Portuguese caravels and the English merchant ships required at least twenty.[5] This advantage enabled the *fluyt* to have lower maintenance costs than ships of rival nations.

The *fluyt* was initially designed without any armaments, and its structure made it somewhat sluggish. The ship was slowly modified, increasing the size of the sails to enhance speed. To protect the ship, the Dutch either armed the ship with several guns or used another modified version of the *fluyt* called a pinnace. The pinnace was square in shape and was built only of oak.[6]

In addition to the development of the *fluyt*, another important innovation that contributed to Dutch maritime success was the wind-driven sawmill. Probably developed in 1594 by Cornelis Corneliszoon, the mill allowed for large amounts of wood to be sawed faster and with substantially less labor than any other nation's sawmills—including the Portuguese and the English.[7] The sawmill allowed for 60 beams to be cut in four to five workdays as opposed to 120 days (the time required to saw them by hand). The sawmills were initially prolific in the Zaan region of the Netherlands. This eventually became the Dutch center for shipbuilding, with about 53 sawmills in 1630. Amsterdam's shift to the use of sawmills was more gradual, but by 1645, the city also had 45 sawmills. The sawmill contributed to the cost advantage that the Dutch enjoyed in shipbuilding relative to its competitor nations. In addition, the net cost of building the *fluyt* with the sawmill was about half the cost of rival nations' ships.

With these innovations, the Dutch could dramatically increase the size of their fleet at lower costs than other nations with more economical trading ships than those of other countries. By the 1670s, the Dutch fleet was bigger than the combined fleets of England, France, Spain, Portugal and Germany.[8] The sheer size of the fleet, the quality of design and the low costs of service as well as the skill of the sailors allowed for Dutch hegemony of the seas by the mid- to late seventeenth century.

The Dutch also created innovations for navigational tools. First, they improved cartographic measures. They incorporated the Mercator projection in their maps—a projection in which lines of latitude drawn on the map have the same length as the equator, allowing for more accurate maps than used before. The Mercator projection allowed the Dutch to create sea atlases, which helped especially with merchant routes that involved sailing across open seas. The atlases were first used in 1598 on the second trip to the East Indies and became standard on voyages of the East India Company.[9] In addition to these atlases, Dutch seamen began using new tables that assisted with determining changes in latitude and longitude. They broke from tradition and began using the sun more regularly to determine latitude, instead of depending solely on the polestar.[10]

Another major innovation developed by the Dutch was the primarily economic limited liability corporation. Before the development of the East India Company (VOC), companies sold shares to investors. The VOC was the first of these companies to introduce the notion of limited liability—the notion that an investor was only liable for the amount that he invested.[11] While not as advanced as the modern-day limited liability corporation (as shareholders had very limited influence over management), the VOC was important in that it provided the finances necessary for the transport of goods and the establishment of political authority at overseas sea ports and colonies. Furthermore, by making company profits the primary focus, the board of directors and the governor-generals that oversaw the board were able to guard against internal corruption.[12]

Drivers of Innovation

How did the Dutch become a global empire within a mere century, advancing from small, partly waterlogged provinces, perennially threatened by the sea? Three institutional drivers played key roles in the improvement of the Dutch: openness, empowerment and competition.

Openness to Immigrants

Openness to immigration played a vital role in the buildup of the Dutch trading empire. Several uprisings in the Spanish South Netherlands (especially the Flemish cities of Bruges and Antwerp) resulted in many prominent businessmen migrating to the northern Netherlands.[13] Due to the inquisitions taking place in Spain and Portugal in the 1550s, many Protestants and Jews fled to the more tolerant Dutch Republic. This influx of immigrants increased in 1609 when the Republic—which covered seven territories, including Holland, Utrecht, Zeeland and Guelders— officially declared a policy of religious toleration. Perhaps as many as 500,000 migrants settled permanently, and an equal number passed through the Netherlands to settle in the Dutch colonies.[14] The population of the nation at that time was about 1.5 million. The Dutch proactively encouraged immigration by providing various incentives to immigrants like tax incentives, start-up capital and reimbursement of relocation costs.[15] Though no common immigrant integration policy was practiced, various cities in the Netherlands competed among themselves to attract the most immigrants.[16]

Abundant job opportunities as well as religious tolerance in the Netherlands encouraged large-scale immigration. Moreover, between 1580 and 1600, many of the immigrants, especially of Jewish descent, brought valuable commercial and technical knowledge and trade networks. The Flemish merchants and artisans emigrating from the southern provinces after 1585 (like the Portuguese Jews) were probably more skilled, richer, better connected and more daring than their Dutch counterparts.[17] Approximately 100,000 South Netherlanders came to the north to flee the inquisition. Many Protestants from elsewhere in mainland Europe fled to the Republic as well. Approximately 30,000 Huguenots (French Protestants) in the later part of the seventeenth century fled France on the revocation of the Edict of Nantes, which had granted limited religious freedom to the French Protestants.

In Amsterdam, which was the most important center of economic activity, foreigners accounted for more than 50 percent of workers in occupations like the seafaring and textile industries.[18] Similarly, immigrants and their children from the southern provinces made up a third of the merchant community between 1580 and 1630. Their wealth was in the same proportion, so that their arrival raised the capital for investments by about 50 percent.[19] The maritime sector was particularly attractive for Norwegians, with 86 percent of the Norwegian immigrants between 1626 and 1715 working as sailors.

Immigrants could also purchase citizenship to enjoy social welfare provisions provided by cities of the Dutch Republic, though sometimes the fee was prohibitive for poor immigrants.[20] *Burgerweeshuis*, an orphanage for children of citizens, was an important welfare institution in Dutch cities. This orphanage was a major support for immigrants who worked as seasonal workers and sailors. Consequently, purchasing citizenship became attractive for immigrants. So, while economic opportunity and religious tolerance led to immigrant influx, the Dutch also had provisions that aided immigrant integration into society.

The massive influx of immigrants was transformative. The amalgamation of immigrants, peoples, ideas, talents and wealth together with newfound independence, tolerance and trade (see next subsection) created a fertile environment for creativity, entrepreneurship and innovation.[21] The Netherlands enjoyed a ferment in terms of these activities. It represented a strong divergence from what occurred in other parts of Europe, which either had traditional dominance by the elites or were engulfed in a paroxysm of inter-religious warfare and persecution. Ideas for the *fluyt*, sawmill, limited liability corporation and other innovations arose from this ferment of creativity, innovation and entrepreneurship.

Empowering Women, Entrepreneurs, Investors and Traders

Empowerment was another key aspect of Dutch society that drove its innovativeness. From the late sixteenth to seventeenth centuries, the Dutch adopted policies and laws that led to empowerment in various spheres of economic and social life in the Republic. These policies and laws transformed the republic into a country of entrepreneurs.[22] Openness to and empowerment of immigrants were key factors in this context, as explained above. The empowerment of women, farmers, property owners, traders and merchants was also very important, as explained below.

Women's empowerment was a hallmark of the Dutch Republic in the seventeenth century. The Dutch seemed very particular about educating girls, especially in arithmetic and merchant accounting.[23] Relative to women in other European countries, Dutch women had more opportunities to participate in commercial activity and consequently played a more important role in Dutch economic life. Indeed, despite Christian insistence on the essential inferiority of women, Dutch society seemed markedly less misogynistic than the rest of Europe at that time.[24] Men did not worry about business continuity after their deaths because their wives could carry on in some of their trades. In fact, Dutch women enjoyed relatively higher economic freedom than those in other regions of Europe during the Dutch Golden Age.[25] Even though women in other European countries (especially England) also actively participated in business and commerce during that time, Dutch women enjoyed more legal autonomy than women in England.[26] Inheritance law was quite egalitarian in the Netherlands, unlike in other European regions. Various legal provisions gave women more economic leeway.[27] As a result, more women contributed to the productivity and growth in the Netherlands than in other countries of Europe during this period.

Property rights in the Netherlands underwent a major change in the sixteenth century, empowering owners, encouraging leasing and fostering trade in lands. Before the sixteenth century, land ownership rights were very clearly established, but lease rights were not. In many parts of the Netherlands, peasants mainly owned land, and the holdings were small in size.[28] Due to clear land titles, the land market was quite transparent and favorable in the Netherlands compared to other parts of Western Europe.

The clear rights to land in the Netherlands ensured that peasants were not at the mercy of a feudal system. In addition, the transparency in property rights and the absence of onerous taxation in the Netherlands compared to England and other northwestern European countries facilitated the easy buying and selling of land.[29] From the fifteenth century onward, land was often sold in public auctions held in churches.[30] These liberal policies toward land in the Netherlands encouraged both increased investment in land for higher productivity and the use of land as collateral in foreign trade.

Before the sixteenth century, lease rights in the Netherlands were not clearly established. During the course of the sixteenth century, however, dramatic changes occurred. In the first half of that century, authorities in the Netherlands started enforcing new rules for leasing land. Written lease contracts with clear termination dates were established,[31] and fines and corporal punishment were stipulated to the violators of these contracts. These policies, along with registering rights to land and houses that started during this time, encouraged landowners to use land as an investment and use the profits for trade.

These changes in property rights played an important role in agricultural development and rural transformation. With clear property rights, there was a phenomenal increase in purchases of land by city dwellers (burghers) and short-term leasing. Leasing was a major source of additional income for the landowners. While land was leased out to tenant farmers, consolidation of land holdings led to economies of scale and more efficient agricultural production. With large land holdings, highly commercialized and specialized agriculture replaced subsistence farming. Farmers shifted their production to dairy, meat and industrial crops such as hemp, which were marketed in cities and beyond.[32] This led to high demand of labor and consequently an increase in wages. Agriculture became more capital intensive than before with investments and innovations in equipment, water management and fertilizers.[33] Improvements in agricultural development had a strong impact on economic growth as a whole.

Simultaneously, with rapid urbanization and population growth, demand for agricultural products—and their prices—increased, giving farmers good returns for their efforts. This change meant that farmers competed for leases, which subsequently led to higher land and lease prices for owners. Large tenants used their high incomes from farming to fund short-term investments in livestock, seeds, tools and labor.[34] Landowners used their holdings as capital investments to generate handsome returns, later used to fund trading activities. Thus, land reforms led to a positive cycle of entrepreneurship, innovation, trade and wealth creation, which fueled further investments in innovation and trade.

The rural transformation impacted maritime trade and shipbuilding in the Dutch Republic. Primarily, the merchants viewed short-term leasing of land as an additional source of income and a basis of creditworthiness.[35] Moreover, an efficient land market with no tax on sales enabled merchants to mortgage or sell their land to fund trade. Thus, the major policy changes that empowered land ownership, development and leasing probably influenced maritime trade commerce and, consequently, innovation in these fields. The favorable policy changes that started in the early 1500s encouraged economic development and prepared the ground for many innovations like the *fluyt* by the 1590s.

Even though favorable systems like land registration and short-term leasing arrangements existed in some parts of Western Europe, the local lords levied taxes on land transactions.[36] Additionally, the transfer of land was complicated, and multiple stakeholders like relatives, neighbors and fellow villagers could claim land ownership. Thus, easy buying and selling of land was not possible in many other parts of Western Europe.

Dutch foreign trade in the mid-sixteenth century was carried out by private merchants working alone or in association with their relatives. Their entrepreneurship led to the formation of various financial instruments, like *partenrederij*, annuities, freight contracts, maritime insurance and promissory notes or private short-term loans.[37] The *partenrederij* was a limited liability contract, first used in shipping but also later in windmills, sawmills, breweries, tile works and paper.[38] It shared the risk of the enterprise to many investors, who had small shares, even as low as 0.016. Loss was limited to the extent of their investments. These instruments enabled even small merchants to invest in trade. Similar mechanisms were absent in England at the same time, where a limited circle of wealthy merchants were key investors. As a result, between 1550 and 1630, while about 3,500 merchants were involved in foreign trade in London,[39] at least 5,000 merchants participated in trade in Amsterdam.[40]

Funding Opportunities and Empowering Individual Investors in the Netherlands versus England

A comparison of Dutch and English investments in trade with Asia reveals the importance of inclusive commercial institutions that thrived in the Netherlands versus England. Prior to 1580, the chartered companies in England could sufficiently invest in trade with support from the Crown. But after 1580, funding was problematic. Merchants had overlapping membership in these companies and were not always able to fund multiple voyages. Often, they were risk averse and waited for dividends from ongoing voyages before investing in new ones. Restrictions in membership prevented the companies from considerably widening the investor net. While English merchants invested only four million guilders between 1601 and 1611, the Dutch invested about twelve million guilders during the same time period.[41] By 1630, the balance of cumulative investments and returns in Asian trade by Dutch investors was five times that of English investors.

The emergence of a money market in Amsterdam after 1595 enhanced the ability and willingness of investors to fund voyages, even before dividends from previous voyages were realized. Investors increased funds by raising short-term loans from merchants and widows, who were hesitant to invest directly in risky trade but were content with a safe 8 percent return on these short-term loans.[42] Additionally, VOC continued to attract investors and had favorable provisions, like payments for share capital in installments and transferability of shares. This development led to an abundant supply of loanable funds and a drop in interest rates.[43] The ability to raise substantial funds helped the VOC establish its stronghold in the East Indies trade, previously dominated by the Portuguese.[44] The Amsterdam money market was a key enabler in this trade growth in Asia. Such a capital market did not develop in England until the 1650s, so English merchants had to pay higher interests to generate funds.[45] Thus, the Dutch were at a considerable advantage in trade expansion compared to the English and Portuguese, due to the highly developed capital markets in the Netherlands.

The method of taxing trade in the Netherlands also played a significant role in supporting trade expansion. From the 1570s onward, Dutch authorities raised revenues from duties on imports and exports to protect the merchant fleet,[46] which encouraged further trade. Moreover, since port cities in the Netherlands were competing for a share of the trade, some of them did not fully impose these duties. In this way, customs revenues from the Netherlands, Zeeland and Friesland provinces increased from one million guilders in 1590 to 2.5 million guilders in 1640.[47] However, in England, customs revenue was treated as a personal income of the Crown, which delegated the responsibility of collection of duties to syndicates of merchants for an annual rent.[48] The custom authorities were therefore incentivized to maximize their personal profits and often abused their positions. As a result, by the mid-seventeenth century, while Dutch trade was two to three times the size of English trade, the burden of customs was two to three times less in the Netherlands than in England.[49]

In summary, the inclusive Dutch commercial regime empowered merchants and farmers to innovate, keep the profits from their innovations and invest in new ventures. Reforms in land ownership, leasing and taxation empowered farmers, landowners and traders to become entrepreneurs. These changes also led to the formation of innovative capital markets that fostered large-scale production and global trade. This shift also triggered innovations in shipbuilding, finance and commerce, which enabled global trade by sea. The flowering of trade enabled further innovations in shipbuilding and financial instruments. This positive feedback loop led to Dutch dominance in European trade in the late sixteenth to mid-seventeenth centuries.

Competition among Entrepreneurs, Cities and Regions

For the Dutch, competition played out at three levels: entrepreneurial, intercity and national.

Entrepreneurial Competition

The Dutch openness to immigrants, the liberal policy toward the making and keeping of profits and the empowerment of enterprise through new land policies and property rights created a considerable class of entrepreneurs. While working in their own self-interest, these entrepreneurs competed intensely with each other. As one historian describes it, the Republic became "a country of entrepreneurs, a society in which the livelihood of a considerable number of men and women depended on their judgmental decisions about the buying and selling of goods and services. These entrepreneurs included not just merchants involved in long distance trade, but also shipmasters, fishermen, millwrights, farmers, artisans, and shopkeepers"[50] in addition to sailors, builders, merchants, traders, financiers, landowners and farmers. Such competition invariably leads to innovations that increase the quality and variety of goods, while decreasing costs. One estimate puts the proportion of entrepreneurs in the Netherlands during the late sixteenth and early seventeenth centuries at about 15 percent.[51] This proportion of entrepreneurship would imply a tremendous level of industrial and agricultural ferment, innovation and growth.

The organization of foreign trade in the Netherlands in comparison to that in England contributed to Dutch ascendancy in maritime trade. In the 1550s, England started exploring new markets like Russia and the Mediterranean by sea, and the Dutch followed suit in the 1580s. However, by the middle of the seventeenth century, the Dutch Republic was more successful than England in foreign trade. In England, chartered companies were formed, which monopolized trade in various European markets. The first known joint-stock company of the modern era, the Company of Merchant Adventurers, based in London, was chartered in 1564. Various influential merchants came together and formed several companies, like the Spanish Company (1577), the Eastland Company (1578), the Levant Company (1592) and the French Company (1609).[52] Membership in these companies was restricted by including prohibitive entry fees, placing a cap on the number of shareholders and making apprenticeships with a company merchant compulsory for merchants who wanted to become members.[53]

In contrast, Dutch foreign trade in the mid-sixteenth century was carried out by private merchants working alone or in association with their relatives. Their entrepreneurship led to the formation of the financial instruments described above. In particular, the creation of the *partenrederij* shared the risk among a large number of investors, who competed to ensure the success of these enterprises and the return of profits. While this practice started initially in trade, it spread to numerous other industries. Similar institutions were not available in other parts of Europe at the time.

Thus, in most markets, Dutch trade had a broad circle of investors from across the Netherlands compared to a select network of merchants in England, mostly based in London. The absence of a trade monopoly in the Netherlands led to competition between companies, which forced them to scout for investors beyond their kith and kin. By then, strong trade links with Asia, Africa and America had been established, and the Dutch voyages reaped handsome profits.

Intercity Competition

In the sixteenth and seventeenth centuries, several coastal cities in and around the Netherlands were in competition with Amsterdam for trade—and for merchants who carried out that trade. These cities included Lisbon, Nantes, Hamburg, Emden, London and especially Bruges and Antwerp.[54] In the mid-sixteenth century, the latter two were larger and richer than Amsterdam. After the Netherlands won independence from Spain in 1586, Amsterdam became an intense rivalry, attracting the trade, merchants and talent from these other cities. To do so, it designed policies that were friendly to entrepreneurs and merchants. In particular, the city was particularly open to immigrants. While other cities also accommodated foreigners, they did restrict them to regions, religions and other classifications. Amsterdam was unique in ensuring that all merchants were equally protected, irrespective of origin, wealth, religion or economic specialization.[55]

As a result of these policies, and with the religious persecutions that existed in other parts of Europe (see Openness subsection above), the population of Amsterdam exploded. Between 1585 and 1609, the merchant community went up from 500 to 1,300, by a factor of almost three.[56] Some of that growth was local. The bulk of it was from the southern provinces. In addition, merchants came in from the northern provinces, Germany, Portugal and England.

One factor responsible for the intercity competition was the relative power of cities to nations.[57] Cities were centers of trade, entrepreneurship and innovation. Nations were cognizant of the wealth-generating power of cities and were thus not eager to tax them endlessly. Thus, cities competed for resources, talent and capital, not only from rural areas but also against each other. With the independence of the Netherlands and the concomitant persecution of religious minorities in other parts of Europe, Amsterdam successfully competed for such talent, resources and capital from other cities to support a class of merchants and entrepreneurs. These people triggered the innovations that led to the Golden Age.

International Competition

Competition at the national level also fostered innovation. The Dutch had two main competitors—the united Spanish/Portuguese empire and the English. Intense rivalry triggered the buildup of Dutch navies and innovations in the design and production of ships. Revenues from success in trade against these nations and the colonization of distant lands provided resources for further naval buildup and innovations.

Initially, Dutch trade revolved around the trade of wheat, wool and fish. Beginning in the 1590s, they expanded trade to include exports of sugar, tea, coffee and tobacco to the Baltic region and beyond into the Mediterranean. In that same decade, the Dutch began trading with West Africa, focusing primarily on gold, ivory and sugar. The Dutch, initially on good terms with the Portuguese, used them as intermediaries with the native African peoples. However, in 1599, the Portuguese king, Philip II, initiated an embargo against the Netherlands in both Spain and Portugal. The embargo pushed Dutch traders to increase their visits to West Africa, expanding from three to four ships a year to twenty ships a year. With their increased visits, the Dutch no longer needed Portuguese intermediaries and were able to dominate the ivory, sugar and most of the gold trade.

Dutch rivalry with the Portuguese was particularly intense in the East Indies. The embargo pushed the Dutch to also increase their voyages to the East. The desire to outcompete the Portuguese, as well as the formation of the East India Trading Company, motivated a military buildup to take over Portuguese outposts in the East. The Dutch established a base in Java in 1619 and drove out the Portuguese from Malacca and Sri Lanka in 1641 and 1658, respectively. In seizing Malacca and the Moluccas (known as the Spice Islands because of the presence of nutmeg and cloves), the Dutch dominated the trade of the most valuable spice in the East: cloves.

The Netherlands' biggest competition was England, beginning at the start of the seventeenth century. The Anglo-Dutch rivalry was initially sparked by the increased Dutch role in the cloth trade and fishing industry. The fishing industry—specifically the herring industry—employed approximately a fifth of the Dutch population and was particularly contentious because the fisheries were off the English coast.[58] The years 1609 to 1621 marked a period in which the Dutch outpaced the British. The British could not keep up with Dutch shipping, freight rates, spice markets and access to silver and a range of other commodities.[59]

The restart of the Dutch-Spanish War in 1621, following 12 years of truce, marked a negative correlation between British and Dutch trade. During the years the Spanish and Dutch were at war, Dutch trade declined, while British trade prospered. Conversely, when the war ended and the embargo was lifted, the Dutch reassumed dominance in the Mediterranean. Dutch trade peaked between 1647 and 1672, coinciding with a time of increased competition and tension with the British. Thus, competition between nations stimulated investments and innovations in ships, shipbuilding and trade so long as the rivalry did not degenerate into outright war. Such war was destructive for all parties.

The competition between merchants and cities sparked innovation in trade and finance. The inter-country competition was more complex: as long as the Dutch were not engaged in active war, the steady state of inter-country competition and the embargo forced the Dutch to increase their participation in trade by cutting out the intermediaries.

Decline

The Dutch dominated trade in the Baltic, the Mediterranean and the Orient for close to a century, from 1570 to around 1670. However, beginning in 1672, the state's trading prowess began to decline. Several factors may have contributed to the decline of the Dutch.

First, success may have sowed the seeds of failure. The country's wealth was increasingly concentrated in a few families. The accumulation of great wealth may have led to the creation of monopolies and cartels that limited entrepreneurship and innovation.[60] Willingness to embrace risk may have given way to risk aversion. The most prominent capitalists invested in government bonds and foreign loans rather than new business enterprises.[61]

Second, mercantilism spread to the rest of Europe. Competition increased among European countries to produce finished goods domestically and promote the sale of such domestic finished goods. The biggest competing nation was England, which developed transformative innovations of its own (see Chapters 9 and 10). Following the lead of England and France, Prussia, Russia and Sweden attempted to reduce the outflow of raw materials and prevent manufactured and semimanufactured goods from entering their countries.[62] For example, Prussia banned the export of raw wool in 1718 and prohibited the import of foreign cloth.

Third, the rise of the British as an industrial economy (see Chapters 9 and 10), further contributed to the decline of the

Dutch empire. In particular, while Britain embraced technological innovation, the Dutch may have been hampered by guilds and other entrenched interests that resisted the new technologies for fear of losing wealth and power.

Fourth, Dutch technology and innovation diffused to neighboring countries in Europe. For example, Dutch processing of raw materials into finished goods, specifically their fine-cloth industry, was the envy of most of Europe. Western Europeans initiated several schemes to harvest both Dutch technology and workers in the industry. In 1668, the French succeeded. By hiring Dutch workers during the Second Anglo-Dutch War and investing large amounts of money, the French were able to increase imports of Spanish wool, increase production of French fine cloth and undermine the sale of similar Dutch cloth in the Mediterranean. This French breakthrough did not entirely undermine Dutch enterprises in the Middle East, but it marked the beginning of other European countries increasing their share of foreign trade. Similar exports of sawing and lumber technology allowed other countries to increase the size of their trade fleets at the same low-cost rate as the Dutch.[63] The spread of technology to process raw materials in other European countries resulted in increased protectionist policies to strengthen domestic production. This policy, in turn, reduced the demand for imported Dutch commodities and contributed to the decline of the Dutch empire.[64]

Fifth, the ability of other European states to build armadas allowed them to increase their presence in foreign trade. These states' absorption of Dutch industrial practices also increased the states' revenue and military capabilities. Together, the states' expanded numbers of ships and forces resulted in the loss of Dutch forts, particularly in West Africa. This loss was evident in the Dutch gold trade, which fell from 484,421 guilders in 1676 to 274,238 in 1702. The Dutch decline in Asia was much slower than that in Europe. However, as the Dutch industrial base in Europe eroded, so did its economic power and trade advantage elsewhere. The eighteenth century marked a gradual decline. Britain's war with the American rebels, whom the Dutch supported, hurt the Dutch. The French further spurred attacks on the Dutch, who provided munitions and transported French and Spanish colonial produce into the New World. Periodic British attacks decimated Dutch trade fleets. War with France, the French victory, occupation in 1795 and the extraction of 100 million guilders (later increased to 230 million) greatly weakened the nation and led to a flight or hoarding of capital.[65] As a result, by 1740, the state transitioned from a major trading power into a simple intermediary trade service.

9

Patenting: Institutionalizing Innovation

In the early fourteenth century, England was open to wool imports from numerous countries. The English then discovered that Flemish wool was of superior quality. Rather than settling for imported Flemish wool, King Edward III decided to improve English wool to compete both locally and abroad with Flemish wool. He granted John Kempe, a Flemish weaver, a "patent" for weaving in the Flemish method—a special privilege, because he was not an English guild member.[1] This "patent" encouraged Kempe to set up shop in England and teach the English his method.[2] However, this was only a primitive form of a "patent." It was granted on a case-by-case basis, without disclosure of the technique to the public and without a guarantee for anyone to apply and receive a patent.[3] Yet, this example shows how openness to outside trade and competition against the Flemish prompted Edward III to grant this primitive patent.

Many such primitive patents did include monopolies, such as the offer of the Duke of Saxony in 1398 to grant a monopoly to anyone who could devise a method of papermaking.[4] However, they were not *modern* patents in the sense that they did not entail the disclosure inherent to modern patents: granting a limited-time monopoly in exchange for disclosing the specifics of the invention. Not until 1624 did the modern system of patenting emerge in England, triggering first a trickle and then a stream of innovations. In 1790, the United States passed a patent statute that was even more accessible to innovators than that in England.

Consequently, that law spawned the growth of innovations in the United States. Countries that embraced the patent system became hubs of innovation, some of then quickly transformed into global economic and military powers (see Chapters 10 and 11).

Why was modern patenting established in England? Why did it then flourish in the United States? Why did it not develop earlier in other parts of Europe, the Ottoman Empire and China? This chapter addresses these issues. Empowerment, openness and competition were important drivers of the early establishment of the patent system in England and the United States. However, the absence or low level of these same drivers may have caused the late appearance or non-appearance of the patent system in other parts of Europe, the Ottoman Empire and China.

The Nature of Patents

What exactly is a patent? The English word comes from the Latin *litterae patentes*, meaning open letters.[5] With a few distinct exceptions,[6] this is still what a patent is today: a document, open to the public, which grants certain legal rights to the patentee. Historically, the rights varied by time and place. Early on, the patentee could be the first inventor or even the first importer of an innovation or craft. Conversely, a modern-day patent is given to only a novel design. The modern patent includes a description of the innovation that is freely accessible to the public and grants a limited-time monopoly on commercialization to the patentee.[7]

A patent does two almost contradictory actions by allowing inventors to profit exclusively from their innovation for a limited time, while simultaneously enabling society to benefit from its disclosure and commercialization. After the stipulated time, anyone can reproduce and commercialize the innovation. In addition, within that time, anyone can also look at the patent and invent around the patent, so long as they do not infringe on the patent itself.

Patents originally arose in order to encourage innovation in the country or region granting them. The governing bodies did so by giving a monopoly to inventors or importers of various trades and machines. A patent that encourages commercialization of an innovation gives that country a competitive edge in trade or battle. England's rise from an island state to a global empire is due to its invention of the steam engine and other innovations that started the Industrial Revolution (see Chapter 10). But that transformation itself was primarily due to its patent laws, which encouraged innovation.

In the absence of a patent, an innovator would be reluctant to disclose or commercialize his or her invention for fear of being copied by others. The patent system reduced that fear and triggered a burst in the development, commercialization and diffusion of innovations. Cities or nations with a patent system would see innovation flourish; those without would be left behind. Thus, patenting is itself an innovation, not at the product or service level but at the institutional level.

Innovation usually requires considerable time, effort and financial resources. At least until the fifteenth century in most parts of the world, one had to ask permission from the authorities to ply a new trade or create a new machine.[8] So, for the most part, the ruler or government was responsible for actively encouraging innovation. This led to arbitrariness in the granting of patents and kickbacks of payments to the ruler.

Over time, authorities devised two main methods to encourage innovation.[9] The first and older method is grants from the authorities, like a direct financial reward. The second is empowering innovators by awarding them with status and special privileges with regard to the innovation.

Drivers of the Rise of Patenting in England

An important driver of patenting was the empowerment of people born from a competitive power struggle between the ruling elites and the people's representatives in the English parliament. This struggle in England was more intense than in other nations of Europe, leading to a gradual disempowerment of the monarch and empowerment of the people and their representatives in Parliament.

The English throne's ability to grant patents came under scrutiny when James I came to power. James was not as popular as his predecessor, Elizabeth, and ruled in the politically turbulent seventeenth century. At the time, there was a growing appreciation for the rights of the individual to property and the limits of the power of the monarchs.

Particularly, in 1624, the English House of Commons engaged in a competitive struggle for power with the king and the House of Lords. One of the points of contention was the throne's tendency to grant patents to those who were able to pay enough. Aside from offering patents for the introduction of an innovative process or product, the king would also grant patents to existing trades, such as the production of playing cards.[10] These patents established new monopolies at the cost of existing businesses; they empowered a few arbitrary start-ups at the expense of the many established ones. Thus, the English House of Commons—whose members represented the people—decided that a wiser policy was to empower innovators and disempower the monarch.[11]

After several attempts, the House of Commons managed to pass the Statute of Monopolies on May 25, 1624. It repealed all past and future monopolies by the throne, making them illegal. It allowed only future patents for completely novel innovations. The authorities were to grant a patent to only the "true and first inventor" for a limited time of 14 years.[12] However, monarchs continued to abuse the process, while Parliament strove to standardize the process. Ultimately, the process that started in the early seventeenth century would only be complete in the late nineteenth century.[13] Despite its limitations, the 1624 Statute of Monopolies was the start of the formal right to patent in England.[14] The statute empowered innovators at the cost of the monarch. It became a template for other countries of the world. It stimulated innovation first in England, then in the United States, then in other countries of Europe and finally in many other countries.

While historical patents turned patenting into a royal privilege, modern patents practically turned it into a right. The right to patent changed the balance of power between the government and the innovator. Now, the law considered the innovator as the owner of the innovation a priori, and he or she had only to register it. Patenting was one of various laws enacted in England during the seventeenth and eighteenth centuries. Some of these laws empowered individuals vis-à-vis the monarch. The thinking at the time was that empowering the individual was not only morally right but would also strengthen the economy by encouraging individuals to compete in manufacturing and trade, creating a free market.

Fruits of Patenting

The English Statute of Monopolies had five major, long-term effects.

First, it created a culture of free trade and an open economy, over which the monarch would eventually have little power.

Second, it in turn empowered individual innovators by establishing a patent as a legal right rather than a privilege given to few. English patent law provided a reward to innovators through a temporary monopoly with opening the market by mandating disclosure of the patent, allowing invention around the patent and stipulating expiration of the patent.

Third, patenting, in turn, generated competition for fame and social status that triggered further innovation. To be the first to patent meant not only profit from sole commercialization but also fame for the inventor. Patenting also led to a record of the inventors of various innovations. No one knows the first inventor of gunpowder because that invention occurred without a patent (see Chapter 4). Indeed, only a few of the old innovations have a known inventor. However, from the eighteenth century on, dozens and then hundreds of innovations have records of their inventors through the patent system. Some of these inventors became famous and highly impactful.

Fourth, patenting enabled wealth creation and distribution. Prior to the patent system, guilds' monopolies on existing trades resulted in a restriction of existing wealth to members of the guilds. Such monopolies perpetuate an existing business class. In contrast, patenting encourages the creation of new wealth. Patenting incentivizes innovators, facilitates social mobility and encourages an entrepreneurial culture.[15] The Statute of Monopolies encouraged not only innovation but also innovative business practices around it. This innovative culture was a considerable force that propelled first England and then the United States into world powers through innovation, wealth creation and profitable trade.

Most importantly, during the eighteenth century, patenting played an important role in stimulating a surge of innovations in England. Three of the most prominent innovations of the period were patented: James Watt's steam engine (see Chapter 10), Richard Arkwright's water frame and Josiah Wedgewood's china. Many of the patented innovations were initially in the up-and-coming English textile industry.[16] As such, patenting was a major cause of the technological revolution in this industry. This had an important impact on the birth of the Industrial Revolution in England, which greatly contributed to England's growth in wealth and power in Europe and, eventually, the world (see Chapter 10).[17]

England's power continued to grow because of the Industrial Revolution, largely spawned by patents. Because of England's prominence and the success of its patent system, various European countries emulated the English patent system. England also spread the idea of patenting throughout its many colonies, the most influential of which would become the United States.

However, while the act of patenting grew steadily in England between the late seventeenth and the early nineteenth century, wealth still determined who could patent because patenting was expensive. Formally, one had to pay £70 for the patent, no small amount in those days. In addition, there were other unofficial costs, such as tips to the officials and the costs of staying in London (if not a local), applying for the patent and waiting to receive the patent (which took two to three months). For this reason, most patentees were well off—if not outright wealthy—and many were from London.[18] Patenting was also a cumbersome process, which further limited the number of patents and patentees, preventing large segments of the population from patenting and contributing to innovation.

Another country saw the limitations of this system and attempted to overcome them. This country was a former colony of England: the United States of America.

Absence of Patenting in Chinese and Ottoman Empires

If patents were so beneficial, why did patent systems not develop all over the world in the seventeenth and eighteenth centuries, as they did in Britain and later in the United States? This question is especially pertinent in places like China or the Muslim world, which, by the mid-fifteenth century, were ahead of Europe in terms of innovation. Indeed, some of Europe's fifteenth-century success, especially with gunpowder, was possible because of the adoption, adaptation or reinvention of innovations that had existed in China and the Ottoman Empire years and even centuries before (e.g., see Chapters 4 and 5). China and the Ottoman Empire had forms of government approval for commercializing innovations, which were similar to early European monopolies, so that certain innovations could not be reproduced without the government's consent.[19] Yet, a proper patenting system in China and the Ottoman Empire did not develop until the twentieth century. Why? The peak of each of the empires (China around the mid-fifteenth century and the Ottoman Empire around the mid-sixteenth century) led to the establishment of dominant all-powerful rulers who faced little internal or external competition, tightly controlled openness and suppressed the empowerment of their peoples.

Minimal Competition in the Contemporaneous Chinese and Ottoman Empires

During the Middle Ages until the mid-fifteenth century, the Muslim world was split into regional powers with considerable competition for superiority. Then, innovation flourished—especially in gunpowder weapons (see Chapter 4). Due to the threats faced by early emperors in the Ming dynasty (1360 to 1431), innovation flourished in China, especially in gunpowder and navigation (see Chapters 4 and 5). However, from their empires' peaks in the mid-fifteenth century, both the Chinese emperor and the Ottoman sultan ruled large empires. Within the empires, there was little dissent or challenge to the emperor's authority. Outside the empires, with a few exceptions, there was little competition from external powers—at least until the nineteenth century.[20]

In contrast, prior to the eighteenth century, England experienced intense rivalry with other European states, though no one country dominated others for any extended period. This environment fostered competition for superiority in trade and consequent competition among innovators. Offering patents was a way for England to attract innovators, produce superior goods and gain a competitive edge in trade.

Trade was important. England was a major trading country before it established the patent law. Its commerce depended on imports for goods, often from far away. England imported staples such as textiles (before its own industry took off), timber and food products. Moreover, British shipping was in constant competition with French, Spanish and Dutch shipping to carry that trade. In addition, the physical proximity of autonomous European neighbors, with culture, language and religions that had much in common, led to the migration of traders, manufacturers and innovators to the most favorable state. This situation created an environment of one state "stealing" a trade secret or an innovator from another. This climate of intense rivalry was one of the drivers that encouraged the rise of patents in England.

In contrast, the Ottoman and Chinese empires covered large landmasses that were highly self-sufficient. China did not depend on much from outside its borders. The Ottoman Empire was involved in trade between the East and West. However, like the Chinese, it largely controlled the main routes in and out of its realms. This allowed the rulers to scrutinize both imports and exports and the entire national economy. Moreover, China and the Ottoman Empire's main imports were luxuries rather than staples.[21] While these empires might also have had areas that possessed regional expertise in a specific trade, they were all under the same rule. So, the dominance of the Chinese and Ottoman rulers over their empires, trade and economy led to little competition. Moreover, relative geographic separation meant their ability to attract foreign innovators was limited—albeit not completely absent—had they felt the need to do so. In other words, inter-country competition was critical to the development of patents in England. In contrast, limited competition did not lead to such a situation in the Chinese and Ottoman empires.

Minimal Empowerment in the Contemporaneous Chinese and Ottoman Empires

Chinese and Ottoman emperors at the time experienced little need to empower their peoples. British governments, even in their heyday, had little in the way of concentrated power vis-à-vis their population. Other political forces such as the nobility or the guilds were quite empowered in Europe. The royalty never quite managed to tame them, even though it tried. A common early use of patents was to infringe on the monopoly power of the guilds. The English patent law originated in a struggle between the House of Commons and the throne, as explained earlier. Various institutions (e.g., royalty, nobility) tried to garner public approval and support by empowering the public rather than allowing other institutions to have it. The throne empowered individuals at the expense of the guild; the nobility empowered individuals at the expense of the throne. Thus, this particular situation of competition drove empowerment, and empowerment drove innovation.

Neither the Chinese nor the Ottoman empire had the need or desire to empower their subjects. They had strong single consolidated centers of power.[22] The Chinese government often hindered the adoption of innovation by preventing those involved from making a profit. The government often monopolized lucrative trade or outright prohibited it. For instance, the Ming dynasty prohibited all maritime trade after 1430 (see Chapter 5). Even when trade was permitted, the government severely controlled pricing.[23] The Ottoman sultans often opposed institutional innovations that would have increased the ability of groups or individuals to make a profit because they had reason to fear the power such institutions might accumulate.

While scholars point out that, theoretically, some institutions, such as the guilds or the Wakf, may have competed for power with the sultans, laws and social norms prevented them from ever being able to take full advantage of this situation. The sultans usually made sure that the situation remained that way.[24] Thus, overall, the Ottoman and Chinese rulers had little incentive to empower innovators by giving them monopolies or patents, and many reasons not to do so.

Minimal Openness in the Contemporaneous Chinese and Ottoman Empires

The Chinese and Ottoman empires at the time were both traditional. Both empires had strong central authorities opposed to novelty. The patent system encourages individuals to invent anything they see fit. Historically, traditional societies are suspicious of novelty, including the Chinese culture, which venerates the past and dismisses novelty as valueless. Rulers that value tradition, either religious or cultural, tend to base their authority on it. These rulers are concerned that innovation will damage their power.[25] For this reason, innovations developed in China were slow to be commercialized.[26] For a brief period in China, from 1360 to 1430, under Emperor Zhu Di and one of his successors, there was an opening to the outside world with seven great naval expeditions to East Asia, South Asia, the Middle East and West Africa; correspondingly, there was a burst of innovation, especially in navigation (see Chapter 5). However, in 1431 the nation abruptly closed in on itself. Consequently, from the fifteenth to the nineteenth century, Chinese rulers mostly treated the outside world and novelty with wariness and were slow to adopt new technology.

Particularly, Chinese closed-mindedness was partly driven by concern that external exposure would harm the existing social order, in which the emperor and mandarins had power.[27] Chinese leadership in silk, china, navigation and gunpowder in the early fifteenth century exacerbated traditional contempt for the civilizations outside the empire, which the Chinese considered barbarians. This view was so powerful that even when innovation did take place in China, it was often couched in language that made it seem as though it were not really new. Such an attitude strongly conflicted with the development of a patent system, which not only rewarded innovators but also honored the idea of innovation.[28]

After the Ottoman Empire's ascendance to power in the fifteenth century, the power of the sultan became absolute, and the empire gradually closed to novelty and innovation. The reason for this lack of openness during this time was the difficulty of achieving a balance between internal stability and external insecurity. There were few internal institutional struggles for power, and the authority of the sultan was absolute.[29] The power of his authority rivaled that of the Chinese emperor. Thus, there was no internal questioning and debate in the empire. During this period, the attitude toward innovation was similar to that of Chinese tradition: it was limited only to instances where it did not challenge the social structure.[30] In other words, when a single authority is dominant in its internal power and not competing with other internal institutions, that authority tends to be less open to novelty.[31] The authority is content with the existing situation, which empowers it and no one else, and does not welcome change that may threaten its power.[32] This was true in both China and the Ottoman Empire but not in England.

Both the Ottomans and the Chinese (with the exceptions mentioned earlier) became hostile to the outside world and to its ideas for different reasons. While the Chinese seem to have truly believed that the outside world had no valuable knowledge to offer, the Ottoman Empire seems to have feared the outside culture. The Ottoman Empire did war with some European countries such as the Austro-Hungarian Empire. This competitive state contributed to a narrow aspect of Ottoman innovation in military technology. Still, because the competition was primarily military and openness to external ideas and the empowerment of innovators was limited, competition did not fully blossom into an atmosphere of innovation. In fact, it was quite the opposite: Ottoman hostility and fear of European culture inhibited any serious cross-cultural diffusion of ideas.

Therefore, with a lack of openness, a patent system did not develop in the Chinese or Ottoman empires. Notably, China did not have a patent law until 1984, when the government started loosening its hold on the economy and the population, and opened up to ideas and innovations from the West.[33]

Absence of Patenting in Florence and Venice

In 1421, Filippo Brunelleschi, a renowned Florentine Renaissance man, designed a new type of ship—one that he promised would make shipping merchandise transport up and down a river faster and cheaper than ever before. At the time, only small boats could sail the river during part of the year. Brunelleschi's invention promised to change all that, helping many tradesmen who relied on transporting materials up the river. However, Brunelleschi "refused to make such a machine available to the public, in order that the fruit of his genius and skill may not be reaped by another without his will."[34] The Florentine authorities decided to give Brunelleschi a three-year monopoly on his invention, so that he would both create the ship and disclose the method for creating it. After the three years, anyone could make his or her own Brunelleschi ship. Brunelleschi's invention held great promise for river trade. Unfortunately, the ship sank on its maiden voyage, much to the chagrin of Brunelleschi and his investors who lost a shipment of valuable marble destined for the dome of the Cathedral of Florence.[35]

Despite this unfortunate incident, Brunelleschi's agreement with the city-state of Florence may have been one of the earliest *modern* patents.[36] The agreement had elements of a modern patent, including a limited-time monopoly given to the inventor in exchange for disclosing the technical construction, so that both the inventor and society might benefit from the invention. Apparently, the Florentine authorities realized that the patent would empower the inventor to further innovate, "so that he may be animated more fervently to even higher pursuits."[37] The patent would also benefit the public by providing access to the new innovation, spurring work for its improvement. The patent would also incentivize other entrepreneurs to generate innovation.

Florentine authorities were not in the practice of making such agreements. Why, then, were the Florentine authorities willing to grant Brunelleschi a patent for this specific invention? Florence had recently conquered Pisa, which connected to the city through the river Arno, and a new ship would help them take advantage of the recently conquered city.[38] This innovation was vitally important, as Pisa had a central seaport that would allow Florence to connect with the outside world. More efficient river trade would boost overall trade, making Florence a more open economy. Florence was also in commercial and military competition with other city-states in Italy, especially Milan, which at the time sought to subjugate parts of Florence.

A decade before the Brunelleschi case, Florence "stole" one Guerinus De Mera from Milan by offering him tax exemptions and a limited-time monopoly for importing the art of making and mounting steel wire bristles for wool-carding machines.[39] The Florentine authorities rightly feared that Brunelleschi, who was very well known in his day, would be "stolen" away by another city-state. In addition, such a ship presented obvious potential economic benefits to the city.[40] That is, competition was a strong motivator for Florence to grant Brunelleschi a patent. Brunelleschi was a notable architect of his day; his patent was likely to have been the talk of the town around Florence and its neighbors.[41] While this patent may not have been all that he hoped for, Brunelleschi did continue to create in Florence for the rest of his life.

Despite that, Florence would not develop a patent system for quite some time, mostly because of the strength of the guilds and the rise of the Medici family. The latter regulated the economy tightly, closing it to free innovation, to various degrees.[42] Similar to Venice, the Republic of Florence was somewhat of an oligarchy of ruling families, though not ruled by one all-powerful monarch.[43] It was in the best interests of the oligarchy to oppose empowerment of the individual, which hindered the oligarchy's financial gain. This alone was a strong impediment to developing a patent system in Florence or Venice. In fact, Italian city-states, as well as France, had much less empowerment of people than England due the power of the ruling elites. In contrast, Britain was fraught with internal power struggles among Parliament, the guilds and the Crown. As explained above, such struggles gave rise to the patent system.

Decline of Britain versus Rise of the United States

The Congress shall have Power [...] To promote the Progress of Science and useful Arts, by securing for limited Times to Authors and Inventors the exclusive Right to their respective Writings and Discoveries.[44]

These words, which the Founding Fathers enshrined in the American Constitution, embody the attitude of the United States toward innovation and patenting. From the very beginning, Congress's fundamental duty was to promote innovation. Where did this attitude come from?

Unlike the English, the American colonists were not entitled to common law. The colonists had no right to patent, but rather, they were dependent on the whims of the British Crown. While each of the colonies had its own patenting customs, the law did not protect these customs from the Crown. Moreover, local authorities rarely enforced these customs.[45] The new constitution following the American Revolution remedied this lack of legal protection from the authorities. So, the Founding Fathers granted Congress the power to legislate a federal patent law, with the wording above.[46] This led to the enactment of the US patent law in 1790.

Due to its origins, patenting in the United States differed from that in England. Many people, including Thomas Jefferson, were opposed to patents; they saw them as illegitimate monopolies.[47] However, others argued that the empowerment of the individual, at the limited expense of the public, was the best way to encourage innovation. The result was a compromise. What could be patented was limited by (a) a close examination to ensure the novelty of the patent and (b) application from only an inventor and not an importer. Thus, the idea of the first and true inventor was extended from the limits of the county to that worldwide. The American government created a patent office in 1802, to examine and grant patents according to law.[48]

Since the idea of economic and social openness drove many of the American laws, the US patent system was more accessible to the public than the English one. Patenting was a lot cheaper in the United States than in England. For example, around 1790, a US patent cost about $4 to $5, while a British patent cost about $585 to $1,680, depending on where it was filed.[49] The price of the US patent went up in 1861, but even then it cost a mere $35. Patenting was also much simpler in the United States than in England. People could apply for a patent from anywhere in the United States, while British inventors could only apply for one in London. The American system also empowered investors: "fake" inventors could no longer take advantage of investors, because now investors could demand remuneration for useless patents. In the American system, patents could be traded: patentees could sell patents easily, and buyers of patents were entitled to their money back if the innovation was of no value.[50] In sum, the American system was superior to the English one in three major respects. First, it was more open than the English system because it was more accessible to inventors anywhere in the country. Second, it was much less costly than the English system (about one-hundredth

Figure 9.1 Global patents from 1790 to 1814

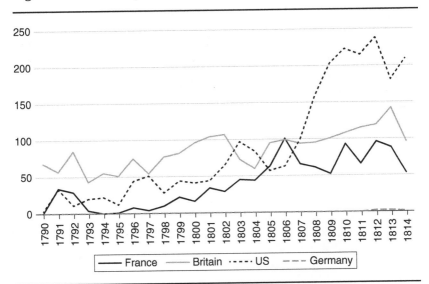

Source: Data drawn from Zorina B. Khan, "An Economic History of Patent Institutions". EH.Net Encyclopedia, edited by Robert Whaples, 2008, accessed November 2017, http://eh.net/encyclopedia/an-economic-history-of-patent-institutions/.

the cost!).[51] Third, it was more competitive because it recognized the *worldwide* first inventor.

The pivotal contribution of the American patent system to innovation in the United States is evident during the nineteenth century, when the country experienced a surge of innovators—people whose main job was to invent. Whereas elsewhere, innovation was either an afterthought or a byproduct, the American patent system—together with greater openness, competitiveness and empowerment—encouraged innovation (see Chapter 11). Companies also began investing in departments of innovation.[52] Consequently, despite joining the Industrial Revolution later than England, the US patenting system became a force that helped the nation become a technological leader. Figure 9.1 shows that only 13 years after establishing its patent law, the United States surpassed Britain in the number of patents issued per year. This is strong evidence in support of the superiority in terms of accessibility of the American patent system relative to that of Britain.

The American system influenced other systems such as the existing English one and the budding Japanese one. The Japanese authorities outright stated that they believed patenting was a main source of American success in the nineteenth century.[53] More importantly, the American system influenced global patent systems.

10

The Steam Engine and the Rise of the British Empire

In the eighteenth century, England broke away from being a European power to becoming a world power. One of the main drivers of this transformation was the innovation of the steam engine. The steam engine revolutionized industry in England and eventually, the world. Horse, water and wind power were popular at the time. But each had shortcomings in terms of reliability, mobility or efficiency. Steam, in contrast, was a reliable, efficient, mobile and abundant source of power. Although initially concentrated near coal-abundant areas, soon the plentiful sources of coal all over the world meant that steam power was widely available. Even today, around 78 percent of the world's energy comes from thermal power, and much of the world's thermal power relies on steam, mostly via steam turbines.[1] Without steam, there would be no steamboats, no steam locomotives, less mechanization and much less electricity—key innovations that led to the development of the modern world. The innovation and development of the steam engine overcame a power bottleneck in mining, manufacturing and transportation that facilitated the Industrial Revolution, first in Britain and then elsewhere.

The fact that British innovators developed the steam engine gave Britain a major advantage. By the nineteenth century—half a century after the invention of the Watt steam engine—Britain was the largest empire in the world. While the steam engine was not the sole innovation responsible for this change, it did play a primary role. Other innovations responsible for Britain's rise were those in the mechanization of production, especially in textile manufacturing and iron smelting. Once the use of the steam engine began in coal mines, other British innovators repurposed it. Inventors like Richard Arkwright, who mechanized the weaving of cotton, and Henry Cort, who developed iron smelting, capitalized on the steam engine.[2] The efficiency of steam power meant their innovations became much more efficient and revolutionary than the prior manual manufacturing methods. These revolutionary innovations drove Britain to become a leading manufacturing power in the global economy in the eighteenth century. Later, using steam for the locomotion of ships and trains drove Britain and other countries that adopted these innovations to further global leadership.

This chapter explores the origin and impact of the steam engine and the drivers that led it to prosper in Britain and not in other countries in Europe or Asia.

Birth of the Steam Engine

The problem of coal became acute in late seventeenth-century Britain. The population was growing and straining the natural resources of Britain, especially fuel. Britain, like much of Europe at the time, relied on coal for fuel. Even before the Industrial Revolution that brought many smoky factories, Britons used coal for household heat, cooking and metallurgy, especially iron smelting. In the late seventeenth century, Britain was running out of easily accessible coal. The coal deposits closest to the surface were growing thin, and coal from deep underground was not easily accessible.[3] So, how could Britain obtain more coal?

The obvious solution for Britain was to mine the coal deeper underground, a challenging activity under the best circumstances. However, in many cases deep underground water flooded the mines, blocking them and causing them to collapse. Mine captains used underground channels to drain mines. This method was both very expensive and severely limited because it required drainable land at a lower altitude than the bottom of the mine.[4] Another option was the use of buckets and manual pumps, using horse or wind power, but these were not always readily available.[5] In the winter, mines were often flooded, causing work delays and serious damage to the mines.[6]

The pressing need for coal made obtaining it invaluable. A burgeoning international market, due to the rise of international trade, meant that whoever could find a way to make mining easy would be highly rewarded. Beginning in the early sixteenth century, Britain established a patent system ahead of all other countries (see Chapter 9). The increasing demand for coal and the establishments of the patent system triggered intense competition for innovation in the mining industry.

Britain was not the only country searching for mining solutions. Denis Papin, a French scientist and innovator in the late seventeenth century, had shown some promise in using the power of steam while working in Britain. The Landgrave[7] of Hesse, a state in the Holy Roman Empire, heard of Papin's achievements and lured him to Germany to perfect and put into practice his ideas for raising mine water with the power of steam. Unfortunately, Papin was not successful in creating a practical machine from his basic ideas. This example reveals the flaw of the patronage system, in which the elite hires people to innovate, and the elite benefits if it chooses the right person, but if it chooses the wrong person, it risks disempowering others who could be successful.

Another example of the failure of high government involvement is the case of Cosimo di Medici II, duke of Florence. In 1641, he hired several engineers to find a method to raise water through steam power. When the engineers were unsuccessful, Cosimo and his government stopped the project.[8] However, an environment that empowers many and encourages competition has a better chance of success than one that does not. The patent system in Britain established such an environment (see Chapter 9).

In contrast to Germany and Florence, the British government empowered the public through the patent system. Unlike elsewhere in Europe, the Ottoman Empire and China, the British economic system encouraged private enterprise much more than that of other European countries (see details below). Instead of the government hiring *one* person to innovate, Britain offered a patent for *any* person to innovate. This system empowered individuals and generated competition among them. Thus, people were free to develop their innovations with minimal government involvement. People were also highly motivated to do so because success would lead to profit and prestige.

The establishment of the patent system in Britain enables us to trace the origin of the steam engine and learn some valuable lessons of innovation in an empowering environment. The first person who obtained a British patent on a steam engine was David Ramsay in 1631, although it appears that nothing consequential came of this patent.[9] In 1663, Edward Somerset, the Marquis of Worcester, was allowed—through an act of parliament—to profit from his steam innovation for 99 years; however, it again appears that nothing came of this innovation.[10] Failures are common in innovations and are sometimes the stepping-stones for future success. Despite these failures, the system of patents with enormous rewards—if successfully commercialized—encouraged many others to strive to earn a patent for an innovative steam engine (see Figure 10.1). In contrast, government patronage, as in Hesse or Florence, can be fatal because failure leaves no opportunity for others to innovate: success must come quickly or not at all. The British patent system, in contrast, allowed the government to continually support innovation because it merely granted a patent for an innovation, and it lost little when innovators failed.

Later, Thomas Savery invented a basic steam-powered pump in 1698, calling it the Miner's Friend. Although it was not quite effective for use in mining (because it could not raise the water high enough), both the Royal Society and the British Parliament clearly saw its potential. As a way of encouraging Savery to develop his idea further, the British Parliament extended his patent protection from the standard 14 years to 21 years.[11] Likewise, many other patents were issued before the steam engine became efficient enough for widespread adoption.

Fruits of Innovation: The Transformative Power of the Steam Engine

The seemingly simple innovation of the steam engine had enormous consequences for the industry, manufacturing, business, trade, growth and global power of Britain.

Figure 10.1 Timeline of the steam engine

1630—David Ramsay's patent for a "fire engine" and other steam applications. It is never put into use.

1663—Edward Somerset, second Marquis of Worcester, receives an act of parliament to profit from his steam innovation for 99 years. However, his innovation is never put into practice.

1690–95—Denis Papin, a Frenchman, invents the steam piston while working in Britain with members of the Royal Society. However, this piston alone is not successful.

1698—Thomas Savery patents his steam-powered pump and publishes the *Miner's Friend*. While the innovation can pump water, it cannot do so to a height that would make it beneficial to miners.

1712—Thomas Newcomen reveals his steam engine, the first engine actually useful in clearing water from mines. Since the engine is not automatic, it requires constant supervision and has a tendency to break down.

1719—Newcomen automates his engine, making it more useful for mining.

1765—James Watt, a Scotsman, invents his separate condenser, thus making the Newcomen engine more efficient.

1773—Watt and Mathew Boulton begin their partnership, creating and installing steam engines.

First, the steam engine overcame a major bottleneck in mining, manufacturing and transportation. It introduced a new form of mechanical power by converting potential energy (in coal) into mechanical energy. Steam is still in use today, in its various forms, as the main way of converting potential energy into power. The steam engine overcame old barriers that limited the exploitation of potential energy.[12] Prior to the steam engine, one of the main sources of power was human and animal muscle. Muscle power has physical limits. Mechanical power, in contrast, can work continuously and indefinitely, as long as one has fuel and the machine does not break down. Steam was not the first mechanical power. Various civilizations have systematically used wind and water power since ancient times. In particular, the Dutch harvested wind and water power extensively. However, both these forces depended on the weather and the location. For example, miners could not use these energy sources, as mines are in limited locations. Also, neither energy source is mobile and both are limited in powering a vehicle. After the innovation of the steam engine, the availability of power was limited only by the ability to acquire coal.[13]

Second, in combination with other innovations, such as the mechanization of weaving and pottery making, the steam engine allowed some countries that embraced it (first Britain, then Europe and then North America) to become world leaders in creating manufactured goods. One of the reasons for Britain's dominance of the cotton industry was the combination of the steam engine and the mechanization of weaving (such as Arkwright's water frame). This allowed the British to compete with Indian cotton, a dominant power in cotton goods for centuries. The mechanization of silk weaving also meant that for the first time, the British could compete with the Chinese, the dominant power in silk goods.[14]

Third, the steam engine transformed the mining industry. It allowed deeper mining not only just for coal but also for other minerals such as tin and iron, essentially increasing their availability. The mining of coal, in turn, increased the availability of iron, as coal was useful in smelting iron.[15] Thus, the steam engine allowed for greater access to metals, which, together with other innovations, changed the face of industry and facilitated the Industrial Revolution. The availability of these metals enabled the manufacturing of a variety of machines, ships and trains, and war tools such as guns and cannons. Britain used these products overseas to conquer and colonize other nations, some of which had not adopted these innovations.

Fourth, innovators adapted the steam engine to ships and locomotives, both of which significantly improved transportation. The steamship made travel by sea more efficient, as the ship was no longer dependent on winds and currents; it made upriver travel easier and turned some great rivers, such as the Mississippi, into major trading routes. Steam locomotives made travel throughout the world quicker. They opened up the American West, in effect contributing to an expedited population growth in that region (see Chapter 11). Overall, the steamship and steam locomotive gave countries that adopted them access to places that were hard and costly to reach, which proved invaluable in terms of trade, farming, mining and other major economic benefits. They facilitated the development of national and international markets, which in turn fueled a new range of innovations.

Fifth, the steam engine revolutionized manufacturing in Britain and later in other countries in a number of ways. It disrupted the British brewing and distilling industries, which were prone to large-scale production and were also labor intensive. For example, in 1787, John Walker patented the mashing device, a machine based on the rotary steam engine that revolutionized the brewing and distilling industries by saving major labor costs.[16] The steam engine provided a new source of power for the cotton and textile industry. Given that this industry required heavy machinery, the new source of energy incentivized the invention of new machines to use this source. As a result, British patents for machine makers went up from 3 percent in the 1750s to 13 percent in the 1790s.[17] Finally, the steam engine required fuel that was expensive relative to windmills and watermills. The relatively high cost of steam energy incentivized innovation in steam engine technology and further improvements in manufacturing.[18]

Drivers of Innovation

The demand for coal was one driver of the development of this innovation; this demand was also prevalent in other countries of Europe. However, during the eighteenth century, Britain was a leader in empowerment, openness and competition, which spurred the development of the steam engine in Britain as opposed to other countries of Europe or Asia. The next three sections discuss these three factors in Britain relative to other countries.

Empowerment through Education and Socioeconomic Climate

The British political, economic and social structure empowered parts of the population more than most other countries at the time. Laws favoring rights of individuals to own property, keep profits from trade and especially monopolize rents from patents greatly encouraged innovation, especially among the emerging middle class. In particular, Britain had the most advanced patent system in the world (see Chapter 9). The British laws (especially patenting) motivated individuals to innovate by allowing them a monopoly on commercializing their patent for a set period of time. The importance of the patent law is evident in that many steam innovators patented their innovations (see Figure 10.1).[19] Thomas Savery and James Watt both managed to get their patents extended and fought hard to protect them, showing the importance innovators placed on profiting from their innovations.[20]

Free grammar schools in Britain facilitated empowerment by educating people. Local landowners often funded these schools to garner the support of the population. In the eighteenth century, a growing class of people who could afford to have their children not work, took advantage of grammar schools. The schools gave these students an opportunity for further education.[21] Both Newcomen and Watt received their primary education in grammar schools.[22] While not every student became an innovator, innovators were more likely to arise from the class of the educated than of the uneducated.

More generally, in eighteenth-century Britain, education became available to more people than ever before. This availability was due to a combination of changes that distinguished Britain from its European counterparts, China and the Ottoman Empire. These changes made education more inclusive and accessible in Britain than in any other part of the world. The changes were supported by the Sunday School movement, which focused on teaching underprivileged children how to read before and after church every Sunday.[23] The result was a major rise in literacy rates beginning in the sixteenth century, so that by the eighteenth century, over half of the British population could read. In Scotland (part of the British Isles), literacy among men was around 90 percent,[24] and in England it was estimated to be over 60 percent for men and 45 percent for women. By contrast, in French- and German-speaking states, it was around 50 percent for men and 30 percent for women. The rest of Europe generally demonstrated lower literacy rates (10 percent to 45 percent).[25] This situation was partly due to, among other things, the availability of schools and reading material.

The government also influenced innovation through tax policy. When Newcomen first came out with his engine in 1712, there was a heavy tax on importing coal. In the 1730s, miners of metals—similar to coal miners—wanted to use Newcomen's engine, but the tax meant that the benefits from the engine were often lost in the cost of coal. Mine owners lobbied to have this tax removed so that they could afford to fuel Newcomen's engine. In 1741, the government agreed, and the use of Newcomen's engine became widespread.[26] Therefore, government action can hinder innovation not only at the early stages of development but also in the later stages of application. The British government understood this and made the use of Newcomen's engine financially viable. In this way, the British government empowered not only innovators but also producers, who contributed to the adoption and diffusion of the innovation.

The success of a given innovation has a sequential empowering effect. Savery's small success encouraged Newcomen's greater success, which ultimately encouraged Watt's greatest success. Thus, the early innovations contributed to the later ones not only through knowledge, but also in encouraging the later innovators to advance and profit from them. Likewise, the innovation of patenting had a considerable empowering effect on innovators across numerous industries.

Aside from the ability to profit, the British socioeconomic climate encouraged enterprise. Unlike in other European countries, and especially in the Ottoman and Chinese empires, being a man of leisure was not the only goal in Britain. Rather, beginning in the seventeenth century and throughout the eighteenth century, industrious business people rose in status relative to the general population.[27] Thus, success in business meant not only financial security but also social status. Unlike in previous periods or other European countries, eighteenth-century Britain opened a narrow path to social mobility, a path that ran through success in business and entrepreneurship. Indeed, the steam innovators of nineteenth-century Britain were popular heroes.

Due to such an empowering environment, Britain became a hotbed for innovation in the seventeenth and eighteenth centuries.

Openness to Scientific Inquiry, Religious Refugees and Social Mobility

During this time, Britain developed a most open environment for research and knowledge, far more than other countries. The open British society attracted philosophers, scientists and innovators from all over Europe. The flourishing scientific communities in Britain exemplified this openness, providing a fertile environment for research, experimentation and innovation. The British grammar schools and education system discussed above contributed to this openness by providing a base of knowledge and creating higher levels of equality among classes of people who learned the same education material. In contrast, in some European countries (e.g., Spain, Portugal, Italy), institutions such as the inquisition still discouraged dissent and open inquiry.

The story of James Watt best exemplifies British openness. When Watt, the best known of the steam engine innovators, finally came up with the solution to the wasted energy of Newcomen's engine in 1765, he claimed it came to him in a moment of inspiration on a Sunday.[28] While the specific solution may have come to Watt in one moment, he had already been working on solving the problem for two years. He had also been working at the University of Glasgow since 1759, when he finished an apprenticeship in London as a mathematical instrument maker. Many years of study and hard work precipitated Watt's seemingly spontaneous burst of inspiration. In fact, Watt only came across the problem when a professor asked him to repair an old Newcomen engine.[29] In other words, Watt's education, life experiences and knowledge were vital factors in his ability to improve the steam engine. All of it was possible in an open environment that tolerated and supported knowledge development.

This situation raises the question of how. How did such a base of knowledge become available to a man from Glasgow who, though not poor, was not rich or of noble birth? Increased education, the rise of scientific communities and the greater availability of books, journals and papers all contributed to the growth of a knowledge base. The advent of the printing press meant that there was more to read at affordable rates so that more people could learn to read and reason than ever before.[30] Consequently, individuals had access to knowledge they would not otherwise have had, such as prior scientific discoveries that contributed to the development of the steam engine. The developments of various innovators of the steam engine—Papin, Savery, Newcomen and Watt—made their way to journals such as the *Philosophical Transactions* of the Royal Society. Other innovators read these journals and published their own work in them.[31]

Aside from literacy, books and journals, other factors contributed to the openness to knowledge in Britain. Improvement of the roads and a rise in the ease and safety of travel meant easier access to knowledge from far away. Technological advancement, a rise in wealth of the families and the Act of Unity (turning England, Scotland and Wales into Great Britain) all promoted travel and business with distant people, especially within Great Britain.[32] The ability to travel had other effects. Not only could knowledge diffuse easily but also skills, ideas and capital could spread as well. Take Watt as an example. Born in Scotland in 1736, he apprenticed in London, returned to work in Glasgow and eventually formed a financial partnership with a businessman and manufacturer from Birmingham.[33]

Another issue that contributed to Britain's economic and social openness was religious toleration. Unlike elsewhere in Europe during the eighteenth century (except in certain areas of the Netherlands), Britain allowed those who were not part of the Anglican Church to live and work in Britain. Catholics did have their rights curtailed, but there was no violent persecution. Protestants, for their part, enjoyed almost complete civil rights even if they were not Anglicans.[34] Ironically, this atmosphere prevailed in Britain during the same time that some Catholic European countries, such as Portugal, Spain and France, were still convulsed in the throes of the inquisition. As late as the eighteenth century, religious dissenters were identified, expelled or burned at the stake in these countries. Such an atmosphere chilled scientific inquiry and progress in these countries. Thus, Protestant and Jewish emigrants fleeing religious persecution in Catholic European countries came to Britain and could use their skills. One such emigrant was Papin, who escaped religious persecution in France. This religious openness also meant that people like Newcomen and Watt, who were not part of the Anglican Church, could live, innovate, profit and gain social status in Britain. Had Britain been less tolerant like other European countries, it might not have been so innovative.

Unlike in Germany, Italy or France, class in Britain gradually grew less important. In many cases, capital became more of a factor than class. For example, in order to obtain a loan and financial backing in Europe, one had to be of the right lineage. However, during this period Britain was developing a new finance system that was based on existing capital and the merit of business ideas.[35]

Both religious tolerance and the gradual increase in the importance of capital over lineage contributed to openness in the sense that a greater number of people could not only innovate but were also willing to invest in innovation. For instance, an up-and-comer like Newcomen, who came from a modest family, could join forces with Savery's business, which was already successful.[36] Another example is Watt and Mathew Boulton's successful partnership—the partnership of a Scottish artisan and a British businessman. While these partnerships seem routine in the modern world, they were actually quite rare in the rest of eighteenth-century Europe because people were more socially, religiously and economically isolated than in Britain.

This openness in Britain enabled more people to innovate. It also facilitated collaboration with investors who had the capital to develop and commercialize the innovation. Due to the patent system and the openness to ideas and immigrants, Britain was filled with innovations at the time, such as those in the energy, mining, cotton, iron and pottery industries. One may also examine the numerous innovations leading up to Watt's steam engine or following it (see Figure 10.1). More importantly, there were not only more innovations but also more innovators than ever before.

Competition among Inventors and Entrepreneurs

The British patent system that empowered innovators and British openness to outsiders and people of different social status created an atmosphere conducive to competition among innovators. To illustrate the competition within the British system, consider the number of innovators who competed to develop the "steam engine" in Britain.

For example, around the same time that Ramsay obtained a patent for his idea of raising water by heat, Somerset, the Marquis of Worcester, was working on a similar idea. Newcomen developed his steam engine (the first engine that was of practical use) around the same time Savery was publishing *The Miner's Friend*.[37] Even though patents prevent competing imitations from entering the market, they do so only for a limited time. Moreover, innovators are always free to invent around the patent and win a patent for their own novel ideas. The success or failure of a patented innovation falls on its acceptance by the market (a vast number of agents), whereas the patronage system relies on the opinion of one individual or a small group of decision makers.

The patent system avoids the politics of patronage because it relies on the merits of the market. Competition for the favor of a monarch is more limiting than competition for the favor of investors and buyers in the market, simply due to the number of people involved. A monarch may favor certain innovators for reasons unrelated to their work, such as for familial connections. However, a large number of possible investors and buyers in the market reduces the influence of such political considerations. The high number of competing potential innovations means that an innovation must be good enough to withstand the competition of the market and the high odds of failure common in innovation. Not only must the innovation work, but it also must work better than its rivals do.

Additionally, the economic policies that contributed to openness across classes in Britain and from immigrants abroad made it a hotbed for internal competition. The limited government involvement meant that innovation and enterprise were open to market forces that favored the successful entrepreneur. In other words, social and economic openness allowed and incentivized people to engage in business and innovation. This, in turn, contributed to a productive, competitive environment, stimulating innovation in Britain more than in other European countries.

Importantly, empowerment and openness in England drove competition between individuals. If not born to a family of nobility, an individual could gain at least some social mobility and wealth through either trade or innovation. This sense of social mobility drove competition between individuals to gain status and wealth through entrepreneurship.

Britain versus the Netherlands and Germany: Why Did the Dutch or Germans Not Develop the Steam Engine?

Until the late eighteenth century (before the age of steam innovations), the Netherlands was England's main commercial competitor. The Netherlands was largely ahead of Britain and other European nations in technological innovation, financial innovation and wealth (see Chapter 8). The Dutch *fluyt* and sawmill, together with many financial innovations, led to the rise of the nation as a premier power in Europe and the world. So, why did the Dutch not develop the steam engine?

Major reasons for the decline of the Dutch in the late eighteenth century were the war with France and the declining level of openness and empowerment. This decline was characterized by excessive control of the guilds,[38] leading to a strengthening of guild regulations and laws in the Netherlands (see Chapter 8).[39] The guilds were business associations of merchants or manufacturers who reached a position of power, wealth and prestige from the status quo. They had little to no benefit from the development of new technologies, such as the steam engine, and generally opposed technological openness.[40] They guarded their own trade methods religiously and tended to be suspicious of new technologies or methods of work because these could cause a loss of jobs, revenue and power. In Britain, many people lost their traditional jobs because of the mechanization of technology. For example, artisans, such as weavers, lost their jobs to Awrkwright's steam and water-powered machine. However, in eighteenth-century Britain, guilds were weak and had little power to prevent the entry of new technology.

The opposite situation occurred in the Netherlands. Guilds were stronger in the Netherlands than in Britain and contributed to the Netherlands' closure to new technologies in the late eighteenth century. Moreover, at the time, the Dutch had a more closed attitude toward technology than the British.[41] The Dutch government did not empower innovators as much as the British did with their patent system and competitive markets. Laws that protected the guilds limited competition. Moreover, the Dutch government at the time was also facing financial duress from war with France and was unable to invest in innovation. While the government occasionally supported individuals, it encouraged them in directions of trade and commerce rather than technological innovation.[42] Additionally, the Dutch had plentiful wind and water power resources, but no major demand for the movable steam engine to drain mines, as much as the British did.

In other countries in Europe, innovation was in control of the government much more than in Britain. In Germany, for instance, the governments tended to control innovation and decide what enterprise was worth the time and effort. Though the German principalities lacked the same extent of empowerment of individuals as Britain, some openness did exist among the principalities. This meant that while a German did not invent the steam engine, the Prussian government was quick to adopt it once it realized the machine's usefulness. The Prussian government then encouraged further innovation, which made Prussia a force to be reckoned with in nineteenth-century industrialization.[43]

Britain versus China: Why Did the Chinese Not Develop the Steam Engine?

While steam could be used as a power source, the only advanced society to utilize this technology was Britain. Other advanced societies either devoted no energy to using steam or lost interest in the idea after first dabbling in it. China, in this context, is particularly instructive.

During the fourteenth and the early fifteenth century, China had undergone a period of economic success, flourishing scholarship and technological innovation, especially in shipping and gunpowder weapons (see Chapters 4 and 5).[44] Unlike the Muslim world and India, China had the basic technology that could have led to an engine similar to that of Newcomen. The Chinese used the double-acting principle (the energy source creates continuous movement rather than alternating between movement and stopping) in contraptions like their water clock that moved continuously and the box bellows that delivered continuous controlled air to stoke fire. The Chinese knew the basic principles that ran a steam engine, as evident by these box bellows, which relied on the power of steam.[45] Therefore, a predecessor to the steam engine may have even existed in China long before it was even created in Britain.

However, the Chinese did not invent or adopt a commercialized steam engine at least until the mid-nineteenth century,[46] under external pressure.[47] The reason for China's failure to develop the steam engine may lie in the stifling environment, especially the lack of progress in competition, empowerment and openness.

First, China lacked the empowerment of individuals and competition among innovators and entrepreneurs. In Britain, competition among individuals drove the effort to build the steam engine. In Britain, if one was not born to a family of nobility, one of very few ways to obtain social mobility, status and wealth was through innovation. China was quite different in this respect. Chinese society respected bureaucrats and attributed more political power to them than to innovators or entrepreneurs. An entire system of study and examinations existed to rise through the bureaucracy. More importantly, and especially during the Ming dynasty (1368–1644), young males from any social status could pursue this path of bureaucratic rise through study and examination. Thus, for the better part of Chinese history, becoming a bureaucrat was one of the only ways to achieve social mobility. Moreover, a career in Chinese bureaucracy could offer political power, social status and respectability more than a military career and much more than a career as a merchant, innovator or entrepreneur. As a result, many ambitious, educated and intelligent individuals in Britain aspired to become innovators and entrepreneurs, while many ambitious, educated and intelligent individuals in China aspired to become bureaucrats.[48] Thus, whereas in Britain, individuals competed with each other in business ventures and technological innovation, in China, they competed in bureaucratic expertise.

Second, empowerment was dramatically different in Britain and China. Only very rarely did some Chinese emperors place an emphasis on technological development (see Chapter 5). However, for most of the time and especially in the later Ming dynasty, the attitude ranged from indifference to outright hostility to technological innovation.[49] Power rested with the emperor and bureaucrats, not innovators. Innovation was primarily an imperial government-sponsored affair, depending on the interests and finances of the imperial sponsors. This meant that even when the government encouraged innovation, the number of people engaging in it was limited to the few the government chose to sponsor. Moreover, after a given number of failures, the government would stop investing in the project.[50] This reliance on royal grants instead of market competition existed in China but not in Britain, as explained above.

Third, while Britain in the eighteenth century was a highly open society, China at the time was a highly closed society. In China, there was a strong pressure to conform to the views of the emperor and the thinking of the bureaucracy. There was little room for divergence, inquiry and dissent. In contrast, Britain had a flourishing academic and scientific community that encouraged openness in views, dissent and inquiry. The Chinese government exam, which was the aspiring bureaucrat's ticket to a government job, required intensive knowledge and memorization of ancient texts. Thus, the best and the brightest of China spent most of their time memorizing classical Confucian texts. In contrast, in Britain, prevailing norms encouraged the best and the brightest to study and develop the newly emerging empirical sciences, engage with scholars abroad and work on novel business enterprises.[51] In Britain, grammar schools were focusing on mathematics as well as language skills, which are useful for business. This emphasis on empirical knowledge allowed the development of machinery such as the steam engine. Newcomen received his knowledge via his apprenticeship, scientific journals and traveling educators.[52]

Watt went to grammar school and learned about engineering and science through journals and his work at Glasgow University. Because Britain highly regarded new knowledge development and exchange, once prototypes of the early steam engine arose, innovative minds from Britain eagerly resorted to replicate and improve on it.

In summary, fifteenth-century China had the basic technological knowledge for the development of the steam engine. However, Chinese individuals did not seize the opportunity, while British individuals did. This difference was due to the significantly varying levels of empowerment of individuals, competition and openness in Britain relative to China.

British Decline

Britain declined relative to the explosive growth of the United States. During most of the early innovations leading to the steam engine, the United States did not exist as a separate entity from the British Empire. After the American War of Independence from Britain (1775–83), the United States started as a separate entity. From its humble beginnings, the United States was strongly committed to openness to ideas and immigrants, empowerment of many of its citizens and intense competition especially among entrepreneurs and firms (see Chapter 11). As a result, during the nineteenth century, the United States grew rapidly in population, innovation and GDP. Early in the nineteenth century, it greatly surpassed Britain in patents (see Chapter 9 Figure 9.1), and innovation flourished in many industries (see Chapter 11). Whereas the young United States depended on Britain for the bulk of its manufactured products, it quickly became a leader in steam technology, industrialization and especially mass production. Americans invented the steamboat in the first half of the nineteenth century, and industrialization in New England was a major factor in that region's wealth.[53] Unlike US farmers, English farmers were extremely late in adopting industrial manufacturing techniques developed in the United States.[54] By the mid-nineteenth century, the United States surpassed Britain in GDP. Thus, prior to

Figure 10.2 Newcomen's steam engine

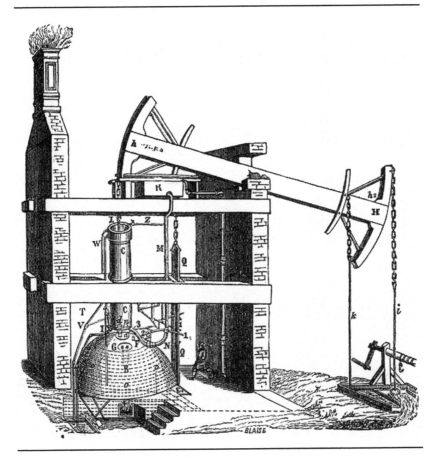

Source: John of Paris/Wikimedia Commons.

the two World Wars of the twentieth century, Britain had already declined as a world power relative to the United States. The growth trend of the United States in population, innovation and GDP continued strongly in the twentieth century, much above that of Britain's. Chapter 11 explores the dynamics of this change and the reasons for it.

11

American Mass Production and the Rise of the United States

At the beginning of the nineteenth century, the United States was similar in economic size to Brazil and Mexico.[1] In fact, Brazil and Mexico may have been more desirable than the United States as destinations for immigrants. However, by the end of the century, the United States had far surpassed Brazil and Mexico. While the US per capita GDP grew six times from 1800 to 1913, Mexico's per capita GDP grew only one and a half times and Brazil's GDP per capita fell.[2] The contrast is even more dramatic in terms of total GDP. Between 1800 and 1900, while the GDP of Mexico and Brazil grew four times, that of the United States grew 45 times (see Figure 11.1). Further, by 1855, in terms of GDP, the United States surpassed the old colonial powers of Europe, including Britain. In 1800, the American industrial output was a sixth of that of Britain. However, it was equal to that of Britain by the late 1880s and 2.3 times that of Britain by 1914.[3] What factors caused the United States' enormous growth relative to Brazil, Mexico and even Britain?

This chapter argues that innovation in the United States was primarily responsible for the explosive growth in the economy. Innovation permeated all aspects of the economy in every industry. What distinguished American innovation from that in Mexico, Brazil and European countries was the thrust for large-scale standardized production (known as mass production) in multiple industries and services. This period of American innovation is

Figure 11.1 GDP of Mexico, Brazil and the United States between 1700 and 1900

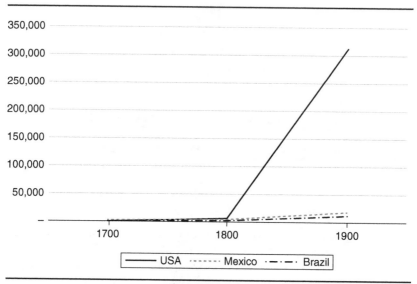

Source: Data drawn from Maddison Project, "Historical Statistics of the World Economy: 1–2008 CE," accessed December 2017, http://www.ggdc.net/maddison/historical_statistics/horizontal-file_02-2010.xls.

sometimes referred to as the American Industrial Revolution.[4] Indeed, this took the Industrial Revolution of England to a whole new level, propelling the United States past England in terms of innovation, industrialization, growth rate and wealth.

After the American War of Independence from Britain (1775–83), there were many doubts whether the newly formed United States of America would survive, let alone be a major world power. Indeed, the young nation had little industry and depended on Britain for most of its manufactured products. However, a climate of intense and continuous innovation permeated every sphere of business and industry, powering an American Industrial Revolution. By the end of the nineteenth century, the American economy had surpassed that of Brazil, Mexico and all European countries—even Britain. The union of 13 struggling states had grown enormously and become a world power. British commentators marveled at the high quality and low cost of American products.

US System of Mass Production

To understand how modern US mass production emerged and became highly innovative, we need to understand how it differed from earlier manufacturing. Scattered examples of standardized manufacturing occurred sporadically in history. For example, the Qin dynasty in China (221–206 BCE) manufactured standardized bronze weapons on a then relatively large scale.[5] The Arsenal in Venice standardized the manufacturing of ships so that they could turn out a ship a day (see Chapter 6). The innovations in Great Britain in the late eighteenth century, such as Arkwright's spinning wheel[6] and Watt's steam engine, ushered in a new era of standardized manufacturing leading to the rise of Britain (see Chapter 10).

However, these cases of standardized manufacturing differed from the new system of mass production in the United States in five distinct ways. First, a tremendous drive arose to minimize human labor and increase automation. Second, a concomitant drive developed to standardize parts and processes in order to reduce costs. Third, the building of massive infrastructure such as roads, railways and canals connected diverse parts of the nation. Fourth, this physical connectedness together with free trade and commerce among the states of the union integrated local markets into the first continent-wide mass market in the world, facilitated by mass distribution and fueled by mass marketing. Such a large mass market could eat up the fruits of automation and standardization, while furthering those forces. Fifth, this system of mass production permeated every sector of manufacturing and service, including agriculture, food, guns, tools, machinery, soap, textiles, railways, shipping and retailing.[7] It involved great efficiencies in manufacturing and distribution, falling prices, rising quality and the design of entirely new products and services. Domestic American innovation permeated every one of these features in every industry.

The United States was not necessarily the single home of major inventions in the nineteenth century. Many inventions still arose in Europe, especially in the first half of the century. However, the United States became the nation of innovation by adopting, adapting and commercializing inventions from all over the world in every aspect of manufacturing, services and business.

Sparse Population as the Initial Trigger for Transformation

Through the eighteenth century, the United States accumulated extensive lands with sparse populations. This situation led to a severe shortage of labor. As such, the United States was in a disadvantaged position compared to the established European powers and the then popular emerging nations of Brazil and Mexico. The first major response to this situation was a burst of openness. The United States opened its borders to unprecedentedly large-scale immigration. Further, those who immigrated were empowered with equal rights to those of prior residents and even given opportunities to farm free land. The openness to immigrants and the empowerment of citizens led to intense competition to make the best use of resources in creative ways, particularly in producing abundant innovations. One particular form of such innovation was the employment of tools and techniques to accomplish tasks mechanically, with minimal labor and high quality, especially after independence from Britain. The result was a unique system of intense mass production that touched every aspect of economic and business life.

Take farming, for example. American farmers employed innovative techniques and machines in the nation's emerging agricultural industries across the United States. For example, by the 1870s, American farmers began using wheat binders and by the 1880s, they employed corn binders.

In Britain and Europe, local farmers did not adopt industrial agricultural techniques as in the United States, until the first decade of the twentieth century.[8] In this case, the gap would seem reasonable, since Great Britain was densely populated and had much less fertile land to industrialize with plenty of labor force to employ for that cause. However, there was no excuse for Brazil and Mexico, countries that had abundant fertile land, as in the United States.

Together with the US agricultural revolution, a group of American entrepreneurs, influenced by the ongoing Industrial Revolution in Great Britain, began to set standards for US textile production. The textile industry at the time developed innovative manufacturing techniques unlike any other place in the world, including Mexico, Brazil and even Great Britain. Some of the most prominent features characterizing the early US textile industry were the large-scale production of medium-quality goods and flow production within vertically integrated mills, seeking efficiency by automation.[9] Vertical integration meant that manufacturing from raw cotton to finished cloth occurred in a single production house.

Why the United States and Not Brazil or Mexico?

Why did such innovation and industry flourish in the United States and not in Brazil or Mexico? We argue that it was due to a unique climate in the United States defined by three drivers. First, empowerment of all citizens (old or new, rich or poor) to own land, manufacture, trade and keep the fruits of their enterprise. Second, open immigration, which attracted people from diverse nations with relatively little regard to religion, economic class or ethnicity, creating a melting pot of peoples, cultures and ideas. Third, intense competition that allowed new innovative entrants to compete with established incumbents. All these forces flowed fundamentally from a creative new constitution, unlike any other in the world. Mexico and Brazil lacked this new freedom and remained countries for oligarchs and elites, who exploited the masses, as will be explained later.

The first driver was openness. The United States had a relatively open policy on immigration coupled with relatively free movement of people within the country. Overall, it welcomed immigrants from all parts of Europe and, to a certain extent, countries of Asia. It encouraged people to move West, stimulating a mixing of peoples, ideas and styles. This constant influx of new peoples, ideas and cultures provided a fertile ground for innovation.

The second driver was an enduring legacy of empowerment. The Founding Fathers guaranteed individual liberties of speech, movement and ownership. During the focal period, new immigrants had equal rights with established citizens. Moreover, new immigrants were not necessarily forced into subservient roles as serfs on the estates of established landed farmers. Rather, the United States had a unique policy to empower landless immigrants. An important law that epitomized this policy was the 1862 Homestead Act that granted free lands to almost any adult willing to start a farm in the western United States, relatively independent of gender, religion, economic background or—to varying extents—country of origin.[10] This act led to a burst of migration westward, the starting up of new farms and intense entrepreneurship, and competition as new farmers strove to survive, succeed and flourish.

The third driver was competition. From the earliest times of the formation of the union, the United States encouraged free competition among businesses. Further, when successful enterprises became monopolies, the United States had a relatively novel policy to cut back such monopolies and encourage competition. We next consider in depth how each of these factors fashioned American mass production. In doing so, we focus on one driver at a time and its impact on one industry, as compared with Brazil and Mexico. In fact, all these factors played out concurrently and fed each, other affecting multiple industries.

Empowerment in the United States versus Disenfranchisement in Brazil and Mexico

By comparing the United States with Brazil and Mexico, we may begin to understand why the attitudes of these three economies toward the empowerment of citizens came to constitute an important feature in the emergence of the innovation of mass production in the United States and not in the other two countries.

Empowerment and the American Agricultural Revolution

"We hold these truths to be self-evident, that all men are created equal, that they are endowed by their Creator with certain unalienable Rights."[11]

These words, which the Founding Fathers enshrined in the unanimous Declaration of Independence of the Thirteen United States of America (July 4, 1776), embody the American attitude toward the rights of individuals. This attitude, which would characterize the United States throughout the nineteenth century, is itself rooted in the unique historical circumstances leading to the English colonial conquest of North America in the late sixteenth century.

England, a minor power during the sixteenth century, entered the colonial scramble for America's territories an entire century after Portugal and Spain had established themselves as world powers. Therefore, when England developed its own colonial aspirations, it had to be satisfied with North America: a backward, less desirable area with then relatively few fertile lands, known precious metals and indigenous populations to exploit.[12]

It took the English nearly a century to realize that what had worked so well for the Spanish colonialists in the South would not work for them. John Smith, a member of the ruling council governing in Jamestown, the first English colony in North America, realized that the colonists would not be able to rely on the coercion of local *indigenous* populations. He asserted that the colonists would have to work themselves or they would starve. The Virginia Company, an English profit-seeking enterprise, attempted to enslave the settlers instead of the local indigenous people. But the effort failed. As a result, it triggered a starving time and subsequent cannibalism during winter 1609–10 due to food shortages.[13] As a result, the Virginia Company adopted a new policy in 1618: it began to base its economic model on incentives that encouraged the colonist settlers to invest and work hard in colonial-mastered projects.[14] This innovation differed dramatically from traditional labor management policy based on coercion that was followed in Spain and Portugal.

Later in the late eighteenth century, the empowering of political institutions in North America led to economic success throughout the British colonies and then the United States. Unlike any other place in the world at the time, North America was the place where empowered and franchised property holders could run the economy themselves. Unlike Brazil, for instance, where royal governors enforced mercantile policies on the colonialists and indigenous populations, in North America, personal interests overtook the interests of the British Crown.[15]

This North American legacy of local empowerment was one of the main factors leading to the American Revolution, at which time the North American colonialists fought for political rights. In fact, the American Revolution against British rule took place much before the independence revolutions in South America and even before the French Revolution and the Napoleonic Wars in Europe, major world events that promoted political rights for non-empowered individuals.

After the independence of the United States, the 13 founding states that established the early United States as a confederation, carried the "torch of empowerment," gradually disseminating it throughout the nation.[16] Thus, in the United States, the first Congress considered the abundant land as the property of the masses rather than that of a few oligarchs or the government, as in Brazil, Mexico and most countries of the Old World. The Articles of Confederation of 1777 maintained that the US Congress would be denied the right to raise tax revenue directly from individuals, though it could from local governments. The Land Ordinance of 1785 allowed the US Congress to bypass the restriction of the Articles of Confederation in a creative way. It allowed the United States to raise funds from the vast lands that it owned by selling parcels to individuals. Such sales empowered people, as they were not restricted on the basis of religion, ethnicity, economic class and, to some extent, country of origin.

The following Northwest Ordinance of July 13, 1787, continued and expanded the empowering policy on land distribution that encouraged individuals to own land in the nation's newly coquetted territories. The 1787 Ordinance maintained that new lands, north of the Ohio River, west of the Mississippi Rivers and south of the Great Lakes, would become part of the United States. The statute determined that slavery would be outlawed within these new territories and protected private property in these lands.[17]

The ordinances of the late eighteenth century were the harbinger of a series of legislative activities that granted the citizens of the United States abundant land that they could exploit with innovative enterprises. Perhaps the most historically momentous act of US policy to encourage individuals to own land was the Homestead Act of May 1862. The 1862 Act granted hundreds of acres of land to individuals—both men and women above the age of 21—who were willing to go West and start a farm. In return, the government expected homesteaders to settle and cultivate the land. Many individuals, including new immigrants and poor veteran citizens, were eligible to acquire this public property.

The new Homestead Act also had tremendous implications for the ongoing social revolution. Since the Act offered land to both men and women, pioneer, working-class women could escape social constraints and take part in the national labor force. Some women homesteaders were merely reluctant followers of their husbands; together with their new husbands they would launch out to the West, where they could acquire a relatively good economic starting point for their new family. However, many women homesteaders were single or widows, who wished to escape restrictive and traditional female-oriented occupations on the East Coast (i.e., maids or factory workers) and find economic and occupational liberty in the West.[18]

The Homestead Act, which exemplified the US policy of empowerment, became a feature of agricultural industrialization in the United States.[19] In fact, starting a farm was a costly business in early America: the owners had to clear, fence and plow the land; acquire livestock, draft animals, tools and weapons; and protect the land against intruders and thieves.[20] However, between 1863 and 1939, some three million individuals applied for land in compliance with the 1862 Act (out of which 5 to 10 percent were women).[21] Altogether, the federal government granted some 264 million acres to 1.5 million households during the period between 1863 and 1939.[22]

The result of these empowering initiatives was the creation of a class of entrepreneurial individuals who were motivated to start and exploit farms through innovation. For example, the number of new farms founded in the state of California soared from 872 in 1850 to about 24,000 over the following two decades. The words of a popular song of the day were "Uncle Sam is rich enough to give us all a farm," symbolically marking the atmosphere of empowerment of the landless masses.[23] Against the backdrop of empowering laws, this ongoing westward migration created numerous entrepreneurs and encouraged them to compete with each other. The result was the more efficient production of livestock, grain, fruits and vegetables of higher quality than ever before.

Henry Miller and Charles Lux, two German immigrants, serve as good examples of the way empowered immigrants innovated in animal husbandry and industrial agriculture after their arrival to the West. In 1858, these two immigrants both started competitive butcher businesses in San Francisco. In their quest for efficiency, Miller and Lux soon shifted into cattle husbandry. Their company came to constitute the largest supplier of cattle in California and one of the most innovative in the field. Miller and Lux introduced innovations in irrigation, cattle rising and meatpacking.[24]

Due to the Homestead Act, acreage allocated to crops in California rose from 32,000 to 6.2 million, and the grain industry in California began producing surpluses for exports. By 1870, California became one of the largest wheat producers in the United States. The fast growth was due to the aggressive introduction of innovative industrial techniques into agriculture; entrepreneurial farmers introduced new mechanized methods into the sowing and harvesting of grain.[25]

The US democratic system was still a fledgling one in the nineteenth century: African Americans and Native American minorities did not receive full access to the nation's resources. However, unlike Mexico or Brazil, in the United States, an intense debate raged over the subject of their rights from the beginning of the republic and ended in a costly civil war from 1861 to 1865. This war coincided with the Homestead Act. The victory of the North paved the path for a unified industrial economy in New England, the Middle Atlantic states and the West.[26] After the Civil War, the Southern Homestead Act was passed in June 1866, partly changing the destiny of African Americans, who had not been allowed to own property before the war.[27]

The 1862 Homestead Act, a symbol of individuals' empowerment by the state during the mid-nineteenth century had far-reaching implications for the following century. According to some estimations, a quarter of the adult population in the United States aged 25 and above (in 2005) had homesteaders as ancestors. Their assets set the base for future generations' economic well-being and ability to play a part in the nation's industrialization as entrepreneurs, workers and consumers of mass manufactured goods.[28]

Mexico: Land for Elites

Between the 1500s and 1700s, the land encompassing present-day Mexico was much wealthier than that encompassing the present United States in terms of raw material, fertile land and available labor force. However, during the nineteenth century, while the United States witnessed tremendous innovation and growth, Mexico (as well as other Latin American countries, including Brazil) did not seem to attain similar innovation and growth.

Mexico and the United States had similarly sized agricultural sectors at the beginning of the nineteenth century. In 1800, about 80 percent of the Mexican and American labor force worked in agriculture. However, the gap in agricultural (as well as nonagricultural) productivity between the two nations became evident as the nineteenth century progressed. Mexico produced only half of US production, revealing that its agricultural labor force was incapable of spurring the nation's economic growth.[29]

One explanation for this gap lay in the different approaches toward the empowerment of citizens that the United States and Mexico practiced. In particular, the explanation pertains to the differing policies toward land ownership, deeply rooted in the colonial era and nationhood histories of these two countries.

During the colonial era, the Spanish Crown fostered a hierarchal economic order based on the exploitation of the indigenous populations in Mexico and nearby countries. The Spaniards developed the encomienda system, determining that indigenous populations had to grant the Spanish conquistadors labor services. Subsequently, the Spanish colonial regime transferred great parts of the local indigenous populations into *reduciones*, new towns that allowed the Spaniards to maximize and control the production within the encomienda system. The encomienda system led to other economic institutions that promised conquistadors revenues from agriculture, gold and silver to the Spanish king but not to the laborers themselves, who cultivated the lands or mined the precious metals.

The encomienda system was first challenged by the collapse of the time-honored Spanish monarchy after the invasion of Napoleon Bonaparte in Spain in February of 1808. At the time, Napoleon enforced the Cadiz Constitution on Spain and its American colonies, introducing the idea of popular sovereignty. The downfall of the Spanish Crown started a sequence of events that led to national revolutions across Latin and Central America that differed in nature from the events leading to the independence of the United States.[30]

In Mexico, Miguel Hidalgo, the leader of the Mexican independent movement, fought the Spanish colonialists in 1810, reclaiming rights for the indigenous (including the right to own land). Hidalgo was a mestizo: an interracial person of local indigenous and European ancestries. His central battle was to break down the ethnic-class stratification in his country that promised ownership of land only to the privileged colonialist Creoles. However, these revolutionary ideas did not manage to put an end to the encomienda. Even when revolting against the Spanish colonial system, the new nationalist elites in Mexico found the Cadiz Constitution idea of popular sovereignty threatening to political stability.

Subsequently, when Mexico finally achieved independence in 1821, the leader of an independence movement, Augustin de Iturbide, became imperator himself and sustained the encomienda for his own benefit. During Mexico's nationhood, the encomienda system flourished, as it had for centuries before, because the colonialists and their descendants ranked themselves at the top and the indigenous populations at the bottom of the economic ladder. It was extremely difficult to uproot this feudal economic order, especially because the powerful elites had strong interests in keeping this order for their own benefit.[31]

The modernization of Mexican agriculture gradually took shape against the backdrop of Mexico's nationhood along the late nineteenth century. The haciendas, Mexico's great farms, underwent a process of commercialization and mechanization, and a rural bourgeoisie elite emerged. During the late nineteenth century, Mexican farmers began to use new machines to grind corn and to thresh wheat, even on the relatively poor farms. However, this development lagged behind that in the United

States. In particular, the modernization of Mexican agriculture was based on the imitation of foreign ideas and innovations. In addition, it relied on import techniques and tools rather than on any local innovations.[32]

During the first 90 years of Mexican nationhood (from 1821 to the Mexican Revolution of 1910), when an agrarian uprising broke the feudal system, Mexico experienced constant agrarian insurgencies. Regardless of the mechanization processes, the production capacity within the haciendas still mirrored great inefficiency: much of the yields were left behind in the field, and the labor remained costly in comparison to the same labor in the United States. For example, the costs of wheat harvest on Mexican farms, such as San Juanico, Juriquilla and Agua Azul, ran up to three times that in the United States, even in the eighteenth century.

One of the reasons for the inefficiency of Mexican farms was that most hacienda laborers worked the lands of their masters rather than their own lands. Landlords based the main work methods on the exploitation of wageworkers, some of whom operated new machines. Even when farm owners imported wheat harvesting machines—some of which were very costly—owners still preferred unskilled laborers with sickles because labor costs were lower that way. As a result, innovation, efficiency and quality suffered.[33]

Mexico's nonempowering economic system was not unique. In light of the rapid industrialization during the nineteenth century, income inequality surged among the world's population. Much inequality developed due to socioeconomic differences within societies.[34] However, Mexico was one of the most hierarchical societies at the time, basing its economic prosperity on the banishment of farmers from their lands. In the late nineteenth century, only 843 families controlled some 130,000,000 acres of land.[35]

The old, long-lived pattern of landlords exploiting and enslaving the indigenous populations ultimately led to the Mexican Revolution of 1910. Emilio Zapata, the Revolution's leader, reclaimed the land for the indigenous underdogs. The Revolution ended with the new liberal Constitution of 1911 that—for the first time—abolished the old, hierarchical encomienda system. The

Mexican Constitution of 1917 replaced the encomienda system with a system called ejido, promising land to the landless peasants. These peasants would organize in groups of at least 20 to form a working community. The Constitution also offered to amend the historical injustice by offering indigenous communities, called comunidades, access to their historic lands. In order to protect the indigenous citizens from hegemonic landlords, the Mexican Constitution of 1917 also prohibited non-Mexican citizens from purchasing land or obtaining concessions for mines or waters in what the Constitution called "prohibited zones": within 100 kilometers from the country's borders and 50 kilometers from its coastlines.[36]

However, agricultural productivity ensuing from these empowering systems could never keep up with the productivity in feudal lands, not because these empowering systems hampered innovation in the first place, but because in Mexico they were very restricted and confined to monopolistic traditions. The new Mexican constitution restricted entrepreneurship by prohibiting individuals within the edijal or comunidade systems from selling, leasing or mortgaging the land, so their ownership was often merely a mirage.[37] However, select foreigners could bypass the "prohibited zone" restrictions by acquiring special trusts from local banks (this situation resulted from Mexico's policy on immigration, discussed later). In doing so, the system hampered innovation and entrepreneurship among the landless masses. In other words, the empowering attempts on the part of the Mexican government were very limited in light of Mexico's strong, time-honored feudal system.[38] As a result, innovation was minimal, efficiency suffered and quality did not develop comparably to that in the United States.

Brazil: Land for the Crown and Its Cronies

Brazil and the United States started at roughly a similar level in terms of abundant land in the eighteenth century. However, by 1913, Brazil's economic production fell well behind that of the United States.[39] Similar to Mexico, much of the explanation of the gap between the United States and Brazil may be attributed to the attitude of the two governments toward empowerment. Brazil inherited its empowerment policy as a legacy from its Portuguese colonial rulers.

During the sixteenth century, Brazil's sugar production was based on the labor force of enslaved indigenous populations. So, already during the colonial period, economic hierarchies emerged in Brazil, similar to those of Mexico. In the Brazilian case, the economic hierarchies were, by and large, the outcome of an extreme economic concentration that the Crown fostered. Vital revenues for the Portuguese Crown came from Brazilian industries, including the booming leather and beef industries in the seventeenth century, which spread across the vast Brazilian land. Thus, the Portuguese Crown, a relatively small economic power in eighteenth-century Europe, kept a centralized economy based on the coercion of indigenous populations.[40] Within this political context, the Portuguese Crown fostered the concentration of land in the hands of Brazil's oligarchy. The Crown created a feudal society that enabled only the privileged elite to own land: it distributed conquered territories by *sesmaria* (land grant), a medieval grant for unused land, enacted in Portugal at the end of the fourteenth century to foster agricultural production.[41]

During the nineteenth century, Brazil witnessed the emergence of a large, ever-growing farming sector that enjoined the country's abundant fertile lands. The cultivation of coffee spread across the southeastern part of the country. Cotton, tobacco and rubber cultivation became common in the Northeast.[42] However, even when an agricultural sector emerged to cultivate Brazil's lands, innovation, enterprise and growth were minimal. By the end of the nineteenth century, agricultural productivity stagnated. This was because Brazil's economy largely relied on agricultural cultivations (i.e., sugar, rice, coffee, tobacco and cotton), worked by indigenous slaves with little incentive for productivity and innovation. Conversely, in the United States, agricultural industrialization thrived because free, entrepreneurial laborers—often homesteaders—cultivated their own land.[43]

The political developments leading to Brazil's independence are key factors explaining the lack of empowerment, competition and innovation in its fledging agricultural industry. During the nineteenth century, Brazil witnessed a relatively mild War of Independence. With Napoleon's 1807 invasion in Lisbon, the

Portuguese Crown fled for safety to its timed-honored colony, Brazil. In 1808, the Portuguese regent of Brazil, Dom Pedro, declared Brazilian independence under his dominion. The outcome was a long-living hereditary monarchy in Brazil that worked to maintain the centralized economy of the country until the Liberal Revolution of 1889. To a greater extent than in Mexico, this Brazilian self-serving political system denied individuals from fully participating in the building of their own national economy.[44]

In 1822, shortly before Brazil's independence, the emperor of Brazil suspended the *sesmaria* system to encourage the use of the country's abundant land. As a result, from 1822 to 1850, no formal rules restricted citizens' rights to access the country's land; in fact, the new Brazilian Constitution of 1824 ignored the issue entirely. Thus, when coffee plantations started to spread due to increasing international market demands after 1830, powerful landlords simply imported more slaves and took control of the expanding frontiers. They could do so because the Brazilian system disempowered individual citizens.[45] Slavery remained the primary system that the local landlords used to cultivate their lands. Cheap labor did not incentivize mechanization, modernization and innovation in agriculture.

In 1850, when a new Brazilian land law was legislated, it was only a result of British pressure on the Brazilian Crown to abolish slavery. The "Lei de Terras" (Law Lands) of 1850 stated that any land ownership prior to that year was to be registered and that new land claims from that date onward were legally subjected to purchase. In other words, the law aspired to fight the common pre-1850 phenomenon of land invasion: that owners of Brazilian lands would demarcate their properties in order to allow the Brazilian government to own and then sell the remaining land.[46]

However, the land remained the asset of the land-owning elite with whom the Brazilian politicians maintained good connections. So, in practice, landowners did not register their land. Moreover, while the law promoted an anti-land invasion policy, the political elite (i.e., the government) never truly enforced the policy. The landed elite thus continued to invade national lands on the Brazilian frontiers during the following decades. In doing so, they blocked any attempts of ownership by free individual laborers and immigrants.[47]

As a result of this regressive land policy in Brazil in the nineteenth century, innovation in agriculture was minimal because land was cheap and vast the economy was declining. During the same time, the United States witnessed explosive growth that led to it dwarfing the early giants of the south—Brazil and Mexico—and becoming a world power.

Competition and the US Textile Revolution

During the second half of the nineteenth century and the early twentieth century, Mexico and Brazil were the largest industrial economies of Latin America, where manufacturing cotton goods was their most important industry.[48] Nevertheless, the US textile industry at the time developed innovative manufacturing techniques unlike those in Mexico, Brazil or any other place in the world. What can explain this divergence in innovation between the United States and these two large nations of Latin America? The answer lies in differing policies toward empowerment and competition in the United States, Mexico and Brazil that drove the differing pattern of innovation and economic developments.

Inter-country Competition between the United States and Great Britain

Francis Cabot Lowell was a US businessperson born into one of Boston's wealthy elite families. From 1810 to 1812, he traveled across Great Britain during its Industrial Revolution, spying on textile manufacturing techniques and technologies within British companies. He realized that these groundbreaking techniques would be crucial for his country's industrial independence. So, he memorized what he discovered. Based on this espionage, he established the Boston Manufacturing Company at Waltham. This was the first mill in the United States, within which the entire production process of cotton textiles took place.[49]

Lowell's case was not the only one of industrial espionage. In the mid-eighteenth century, British domination of North America was the main challenge to American independence and national ambitions. Immediately after the American Revolution (1765–83), the United States still relied on the supply of manufactured goods from Great Britain, despite its political independence.[50] At the time, the British still tried to exploit the United States as a colony. During the Napoleonic Wars from 1803 to 1815, despite US independence, the British navy tried to force American sailors to join their maritime efforts in their combat against France. Later on, in 1812, British and American troops engaged in a military clash on American soil.[51] The emerging ambition of the American political elite to close the gap with Great Britain throughout the nineteenth century developed against the backdrop of military and political rivalries between the nations.

Within the context of political and military tensions, the British ended up restricting the accesses of American sailors to European shores. The result was the US Embargo Act of 1807 on British commodities. The embargo, coupled with the ongoing warfare between the parties, could have been a disaster for the fledging United States. However, ultimately, the embargo boosted innovative industrialization in the United States due to its drive for a more independent, efficient and domestic industrial system.

For instance, during the War of 1812, the army demanded uniforms and tents in large numbers. The demand resolved a long-standing internal debate within the United States over the value of an autonomous industry for national security and development. It was now clear to most US policymakers that the new nation could only rely on its own self-sufficient national industry.[52] Therefore, the 1812 incidents accelerated the industrialization of domestic cotton agriculture; it was mostly cotton fields that supplied the cotton for military tents and uniforms produced by local manufacturers.

The competition between Britain and the United States during the early nineteenth century set the stage for an innovative US textile industry. Technologies of the textile, iron and steel production in the nineteenth century were almost entirely British. Britain was the world's number one exporter of these materials and still the producer of the world's finest steel. British policymakers sought to hide such technologies from the United States by casting a technological embargo and an emigration ban on experts in these fields.[53] The US policymakers, on their part, adopted a strategy that did not recognize British technological patents. US agents searched for skilled artisans by crawling British cotton districts. They offered cash awards to those British experts who were willing to break their country's embargo.[54] US entrepreneurs stole, bought or copied, and adapted knowledge and technology from Britain in whatever way they could.

Intra-country Competition within the United States

Lowell's story illustrates the beginning of the era of industrialization in the United States. Following the War of 1812, a self-sufficient US textile manufacturing system took shape at Waltham and then across New England. The "Waltham model" that the Boston Manufacturing Company created applied simple British technologies to increase the pace of production. Lowell joined two local engineers, and together, they were able to improve British techniques—mostly because they had succeeded in mechanizing the weaving process. Lowell and his partners successfully integrated all the stages of production within one mill. By 1820, the Boston Manufacturing Company produced some 450,000 pounds of cotton using 5,376 spindles and 175 power looms.[55]

These developments occurred within a clear US policy, since there was an independence of free competition among entrepreneurs. A prominent early instance was the so-called Jeffersonian Democracy spirit (named after Thomas Jefferson). It referred to a group of Republican politicians who, during the late 1700s and early 1800s, encouraged economic mobility among grassroots "yeomen" farmers. These politicians denounced established economic elites.[56] The Jeffersonian Democracy marked the harbinger of a procompetitive attitude among US policymakers.

Within this political context, the quest for efficiency among early US manufacturers, like Lowell and his partners, led them to seek industrial areas where British imports would not constitute clear competition. For instance, a group of New England textile manufacturers found out that low-quality fabrics including cassimere, kerseymeres, satinets, cassinets and flannels would meet the demands of the military and the slowly expanding civil mass markets. Manufacturers were able to reduce the costs of production of materials such as cassimere and kerseymeres that, unlike broadcloths, did not demand much refinement during the process of their production. Flannels could simply be washed and ironed to attain their final appearance. Lowell's Boston Manufacturing Company confined its production to standardized cheap cloths, sheets and shirts made from such fabrics. Another innovation that the Boston Manufacturing Company presented was the organization of the workforce. Lowell and his partners housed young women from New England farms in a workers' area near the mill.[57]

The impetus for innovation in Lowell's company and others in New England was not only imported British textiles but also the fledging textile industry in neighboring Rhode Island. Manufacturers in Rhode Island, which sprang up as an outgrowth of New England's textile industry, resorted to forms of small-scale quality productions based on the labor of family units. These units used better manufacturing methods than those the British immigrants had brought to the region a few decades earlier.

Industrial innovation, particularly within vast unpopulated lands, needed infrastructure. As early as the beginning of the nineteenth century, competitive and empowered entrepreneurial farmers in New England started investing in canals and bridges, which they saw as efficient means of transportation for their own goods and economic liberty.[58] Together with traders, they launched distribution businesses across New England, all the way down the Hudson River and beyond—reaching as far as New Orleans in the South.[59] Because such distribution infrastructure unified the fledging US industry, innovators and businessmen who sought to invest in them could acquire financial support from the federal government. The government launched joint public-private companies that allowed entrepreneurial farmers or traders to contribute to the ongoing transportation revolution.[60]

The construction of the Erie Canal brought about further competition among entrepreneurs in the nearby areas. For instance, the construction of the Erie Canal set the stage for innovative industries. Some competitive entrepreneurs in the United States who strove for efficiency were willing to use the new transportation infrastructures to travel far distances in search of new markets. An example is the manufacturing and distribution of stoves for domestic use. Because stoves were rather heavy, distribution by peddlers was ineffective. With the 1825 opening of the Erie Canal, stove manufacturers began to distribute their products more efficiently from distribution centers such as Albany or Tory.[61]

The spirit of empowerment in the United States led to the development of a patent system modeled on the one in Britain, but much less expensive and more accessible (see Chapter 9). The patent system triggered competition among entrepreneurial individuals who aspired to become the first inventors in their field. As a result, the US patent system triggered a burst of innovation across industries that was greater than that in the Industrial Revolution in Britain (see Chapter 9 and Figure 11.2).

In the Americas, the same economic system that generated competition among entrepreneurs sought to break up successful firms that gathered too much power and grew to be monopolies.

Figure 11.2   Historical patents among major world economies

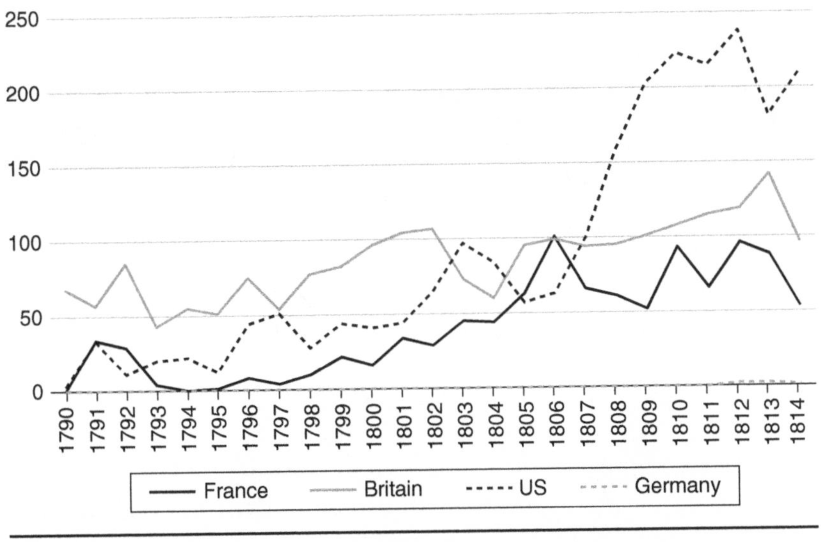

Source: Data drawn from Zorina B. Khan, "An Economic History of Patent Institutions". EH.Net Encyclopedia, edited by Robert Whaples, 2008, accessed November 2017, http://eh.net/encyclopedia/an-economic-history-of-patent-institutions/.

This flexible policy that encouraged start-ups to grow while cutting back monopolies created a vibrant climate for industrial innovation. The policy encouraged entrepreneurial individuals to compete with each other throughout their attempts to attain great economic success. At the same time, it ensured that future entrepreneurs would be protected from unfair competition by successful ones who had become dominant.[62]

Entrepreneurship and competition were further facilitated by the presence of a vibrant banking system in the United States. Competitive markets can truly emerge only if a large number of entrepreneurs can easily raise funds to grow their innovative ideas into flourishing businesses. In the United States, many entrepreneurs could start small firms and turn them into profitable businesses because of a stable banking system, mercantile houses, companies or individuals that lent them money. In 1818, only about 338 banks operated in the United States, but by 1914, the number reached 27,864. In fact, these banks helped boost the textile industry because they competed against each other for the business of potential clients. The results were rather low rates of interest for those wishing to start new textile firms.[63]

The combination of empowered individuals, competitive markets and easy access to capital set the stage for great innovation in the US textile industry. Starting in the 1820s, New England's cotton textile mills began to organize in public corporations, which raised public funding. In 1835, some 14 textile firms, including the Boston Company, were traded on the Boston Stock Exchange, and by 1865 there were 40.[64] This was the fruit of empowered entrepreneurs who competed against each other and built up to a network of industries as well as a large banking sector, sustaining each other's growth.

The Boston Manufacturing Company cooperated with other textile companies slowly organizing networks of textile manufacturers. By 1845, the network outgrew: 31 Massachusetts textile companies produced 20 percent of the entire national cotton textile. With their earnings, these textile companies started investing in infrastructures, such as the Boston and Lowell Railroad Company, or in banks that financed their investments. These textile companies controlled 40 percent of the banking capital in Boston and 30 percent of Massachusetts railroads.[65]

Therefore, paradoxically, several monopolies in the United States emerged from ongoing intense competition, as some individuals gained extensive success. However, they were short-lived. Those who managed to compete and who eventually prevailed could become a profitable monopoly, only to be threatened by new entrants that grew on the basis of an innovation and competed for those profits. In other words, monopolies were not protected by a privileged historic elite, but were subject to intense competition from new entrepreneurs. This pattern of innovation and competition has survived in the United States until today and is typical of the environment in Silicon Valley.

To maintain such conditions and sustain fair competition, the US government also sought to break down monopolies in the textile or banking sectors. The National Banking Acts of 1863 and 1864 are good examples. The Acts created a network of chartered banks that would sell government bonds to less fortunate entrepreneurs who had been left out of the network. By doing so, the US government allowed rather easy access to finance that boosted its textile industry.[66]

Thus, political institutions of the United States balanced the spread of economic and political power and reduced the chances of crony capitalism. A politician knew that to be reelected, he or she had to support and promote the aspirations of a large group of individuals. From time to time, politicians restrained the concentration of economic power among economic elites.[67] They cut back monopolies, but still allowed ambitions of wealth to emerge among motivated entrepreneurs, regardless of their relatively weak starting points.

From the 1890s to the 1920s, various movements of social activism like the Populist and Progressive movements and a series of subsequent legislations worked to reduce the power of "trusts." Trusts were large industrial monopolies that sought to formalize the networks system that allowed large corporations, such as the Boston Company, to grow even larger. Active antitrust politicians employed special acts to break up such large monopolies. Teddy Roosevelt, known as the "Trust-Buster," became a great threat to such strong commercial monopolies, specifically after his success in the presidential election in 1901.[68] By governing and restricting the power of the few, the antitrust spirit since the 1890s paved the way for empowering the masses and fostering fair competitive markets. The antitrust spirit was fully institutionalized when Congress passed the Federal Trade Commission Act and the Clayton Act in 1914. These acts set the stage for the founding of the US Bureau of Competition, with the goal of encouraging competition by enforcing antitrust laws.[69] A similar goal was shared by the Anti-Trust Division of the US Department of Justice, which also had the right to file charges against violators of the laws.[70]

The competitive atmosphere in the United States brought further innovation into the fledgling textile industry. By the last third of the nineteenth century, techniques of ready-made garment manufacturing gradually started supplying what were previously considered upper-class luxuries to the masses. A series of economic crises between 1893 and 1907 disseminated the trend of wearing ready-made garments among the middle class.[71] In addition, until the early twentieth century, the ready-made garment industry produced clothing mostly for men. In 1890, the women's garment industry constituted only 25 percent of industrial clothing. High prices made industrialized garments unaffordable. So, skilled garment workers used their expertise to make their own clothing at home after work. It was only in the 1920s that mechanization dramatically reduced prices and made ready-made women's clothing affordable and popular. Several factories produced fashionable dresses for the masses.

The spirit of empowerment and competition that fostered the textile industry in New England did likewise in numerous other industries in the United States, including agriculture, food, guns, tools, machinery, soap, railways, shipping and retailing.[72] This development was in marked contrast to that in Mexico and Brazil.

Mexico and Brazil: Pandering to Monopolies

The development of local textile industries in Mexico and Brazil unfolded quite differently from that in the United States. The former were typically noncompetitive. In the mid-nineteenth century, Mexico lacked an efficient banking system, while a stock market simply did not exist. The Mexican government opened the first Mexican bank in 1830 to support investments in the nation's textile industry. By 1911, the number of banks in the country reached only 47. Moreover, three large banks—Banco Nacional de Mexico, Banco de Londres y Mexico and Banco Internacional Hipotecario—controlled two-thirds of the local banking capital. Because the number of banks was small, competition among Mexican banks was limited and interest rates were high—between 12 percent and 40 percent per year, and sometimes 10 percent a month.

The owners of large enterprises could gain access to capital only through their family and kinship networks. Instead of banks, large merchant houses handled commercial transactions by issuing letters of credit only to a few wealthy businesspersons. Coteries of entrepreneurs and bankers, who belonged to better-endowed networks, eventually eliminated any threat of competition from those who came from a less fortunate background and did not have access to investment fund resources.

In 1896, the first traded companies appeared on Mexico City's stock exchange. By 1908, only fourteen companies were traded, out of which only four were textile cotton manufacturers. The later four textile manufacturers constituted only 3 percent of the entire Mexican textile industry. The situation did not change until 1930. The fact that individual entrepreneurs did not get a chance to invest, compete and earn profits sustained a vicious loop that hampered public investments, innovation and entrepreneurship in the local textile industry. The result was an inefficient textile industry.[73]

Antitrust laws and activities that broke up monopolies in the United States simply did not exist in Mexico. As late as 1930, most of Mexico's textile industry was financed through the same sustaining kinship networks of wealthy businesspersons started a hundred years earlier. The owners of textile companies were connected to the political system through tight interpersonal ties that closed the industry to new entrepreneurs.[74] Most mass production technologies in Mexico were imported into the county. They were concentrated in the hands of a number of foreign-owned firms that relied on foreign innovations.[75]

Moreover, Mexican entrepreneurs could not compete with each other as in the United States because economic elites and the government restricted their physical access to the country's industrial centers. The best land available—mostly in central and southern Mexico—remained in the hands of the local elite of Creoles who had been well established since the colonial era. Some of them were even prestigious religious elite who possessed church land.[76]

Similar to Mexico, the political-economic system in Brazil reduced the ability of an individual to start a firm and compete fairly. Brazil had only 26 banks in 1888, which limited access to the broad public. From 1850 to 1885, only one manufacturing company was traded on the local stock market.[77] Moreover, the Brazilian Crown promoted local monopolies. One of them was the monopoly dominating the transportation on the Amazon River, which the king granted to Baron Mauá (Evangelista Amaral de Souza), Brazil's richest businessman, who had seemingly bribed the king.[78]

In concert, another reason for the limited competition over Brazil's large plantations in the south, or in the north's Amazon Basin, was that from the beginning of the nineteenth century, the Brazilian Crown had a monopoly over navigation. The economic wealth of rich plantation owners was coupled with enormous political power. Based on their political and economic power, plantation owners were able to persuade the Crown and local state governments to deny the free workers entry to their territories. Thus, potential entrepreneurs who did not have access to slaves could neither obtain the labor of paid workers.[79]

A coast-to-coast monopoly of navigation in the early nineteenth century was followed in 1836 by a law that confined coast-to-coast and any internal navigation to Brazilian-owned ships, meaning that their owner had to be a Brazilian resident and not share ownership with a foreigner. Moreover, the conditions of the law determined that 75 percent of the ship's crew, including the captain, had to be Brazilian. Thus, the already partially monopolized transportation norms (e.g., on the Amazon River) were now further restricted.[80]

In addition, those entrepreneurs who did wish to make the move toward the coffee plantations in southern Brazil or across the Amazon—regardless of the limiting monopolies—had to face poor transportation due to the lack of competitive investors who would invest in infrastructure. Unlike the United States, Brazil's economic system lacked the institutions that enabled free workers to invest in transportation infrastructures or finance their own start-up. As a result, only slaves from within the country or overseas, whose wealthy employers financed their costly transportation, could engage in interregional labor-related migration. Much of Brazil's population, mostly in the north of the country, had to struggle to move across their own country to pursue their dreams or carry out their entrepreneurial ambitions. Instead, landed elites monopolized business and restricted individuals' ability to execute their competitive enterprises.[81]

As a result of this stifling and uncompetitive setting, Brazil in the nineteenth century did not develop an innovative and dynamic manufacturing industry such as the one that flourished in the United States. Development stagnated and the economy even shrunk, whereas it grew sixfold in the United States.

Transformative Immigration: US Openness versus Brazil's and Mexico's Closed-Mindedness

The third driver that furthered the innovation and growth of US mass production was the openness to immigrants, who constituted a vibrant and entrepreneurial labor force. At the founding of the nation in 1776, the United States had a much smaller population than Mexico, Brazil or Britain.

Unlike the newly founded United States of the late eighteenth and early nineteenth centuries, the Victorian British Empire, "on which the sun never sets," could rely on India and other colonies for a constant demand of goods and a steady supply of labor. The United States, like Mexico and Brazil, had abundant resources such as land, water and timber.[82] However, it lacked people to exploit those resources. Moreover, its resources were not concentrated but dispersed over a huge land mass.[83] Yet, of these three nations (Brazil, Mexico and the United States), only the United States broke off to become a superpower. Why so? Besides empowerment and competition discussed above, a third driver was a relatively great openness to immigration in the United States, which did not exist in Brazil and Mexico.

The United States: Open for the World's Workers and Entrepreneurs

A characteristic of openness in the United States was its unique, welcoming immigration policy that—with two major exceptions— was relatively open to people of different ethnicities, religions, countries and cultural backgrounds.[84] One exception was people from Africa, who were allowed into the country as slaves until the end of the Civil War in 1865. The other exception was Chinese, who were barred from the country by the Chinese Exclusion Act of 1882 until its repeal in 1943.

New immigrants were empowered with the same rights as native citizens and quickly assimilated into the population. Openness, equality and empowerment made the United States a "melting pot."[85] The US population increased dramatically from about 2.5 million in the mid-1770s to about 92 million in 1910—more than double the size of Great Britain.[86] In comparison, Mexico's population increased from about 5 million to 14 million.[87] While high birth rates and declining mortality played a significant role, much of this demographic growth was also the outcome of immigration from overseas and their descendants.[88] Between 1820, when the US government first kept track of immigration, and 1920, at least 35 million newcomers flooded the new nation.[89] After the number of their American-born offspring is added to this number, it is evident that immigration contributed substantially to the US population growth.

With the founding of the republic in 1776, US openness embodied a federal policy to maintain open doors to immigrants. The aforementioned 1862 Homestead Act was not only a milestone in US policy on empowerment but also in its openness.[90] The United States promised new economic opportunities and open lands to people from most of the world rather than only to its own citizens. For example, European immigrants experienced considerable upward occupational mobility after their arrival. They were able to enter the farming sector that was blocked to many European immigrants in their native country of birth.[91]

The British government imposed draconian laws on displaced farmers and traditional artisans, whom local elites saw as threats to social and economic order. The British Industrial Revolution gave rise to a new class of urban "free proletariat," but norms of "unfree" labor played an important role in that nation's industrialization. While Britain oppressed such unskilled and displaced British farmers, the United States offered them and other marginalized people a new home. In the United States they could live and work without oppression from traditional values, monopolies and powerful elites. Consequently, British grassroots immigrants moved to America's industrial areas in the nineteenth century. Overall, from 1800 to 1860, 66 percent of the entire immigrant population in the United States came from Britain, even though (and perhaps because) this was when Britain's Industrial Revolution peaked.[92]

Albeit prominent in the overall immigrant population, former British citizens were not the only nationals who provided the United States with fresh human capital needed for its fledgling industry. Though many of the immigrants in the nineteenth century came from Northern Europe, they were relatively diverse, also coming from eastern and southern Europe, Asia and even Mexico.[93]

Besides economic opportunity, the US government promised freedom of religion to those suffering from religious persecution. For instance, the constitutional separation of church and state protected immigrants from religious discrimination. The absence of ethnicity-based restrictions on employment and freedom of worship attracted refugees and migrants from across the world.[94] The largely Protestant society in the United States in the early nineteenth century became increasingly religiously heterogeneous due to the mass immigration of Catholics from southern and eastern Europe and Jews from various parts of Europe.[95]

The US government sought to empower unskilled immigrants and create healthy competition among all. This policy succeeded very well. Between 1840 and 1920, immigrants in the United States earned just as well as natives—even better on certain occasions. Consequently, they could own a home, save and spend just as domestic citizens.[96] This was a key factor that further promoted an equal society and an entrepreneurial spirit.

The availability of a labor force enabled the rapid industrialization of the United States in the nineteenth century. Canal and railroad companies recruited workers directly from Europe. Irish, Italians and Jews from Eastern Europe entered the construction, transportation and industrial sectors in the port cities of the East Coast, and the same was true for Chinese immigrants in the West. Other European groups helped developed the heavy industry of the Midwest.[97]

Against this extraordinary open and empowering policy on immigration, newcomers could pay their own costs of immigration. For example, a credit-ticket system helped Chinese immigrants buy into their immigration to California. The system included employers paying for the worker's trip, with the immigrant worker repaying the debt after sufficient earnings.[98]

In the early nineteenth century, the farm workforce in California comprised mainly European descendants. However, beginning in the mid-nineteenth century, international immigration from other non-European origins rose. Chinese immigrants entered the farming industry around 1857. The Chinese constituted some 10 percent of the entire farm workforce in California.[99] In 1876, the number of Chinese immigrants in the United States was 151,000, of whom 116,000 chose California as their new home. In the mid-nineteenth century, the governor of California, John McDougall, referred to the Chinese newcomer as "one of the most worthy of our newly adopted citizens."[100]

Nonetheless, immigration also had at least two stains in an otherwise remarkably open chapter. First, the United States— especially the Southern states—was home to slavery, though the policy had been intensely debated since the founding of the nation. This debate ultimately led to the Civil War in 1861. Slavery was abolished in all states in 1865, though discrimination persisted for more than a century later. Second, the Chinese Exclusion Act of 1882 banned an entire ethnic group—the Chinese—from immigration because of their race.[101] Immigration law also had restrictions, especially in granting citizenship. Initially, after the founding of the country in 1765, citizenship was limited to whites. From 1865 on, citizenship was extended to slaves and their children (African Americans). Only after the 1960s was citizenship fully extended to Asians.

Nevertheless, in the nineteenth century, the US immigration and citizenship laws were far more progressive than those in Mexico and Brazil. While migration was taking place on a global scale during the nineteenth century, it was unprecedentedly wide and remarkably fluid in the United States. From the mid-nineteenth century, the United States opened its borders to millions of newcomers, who turned it into the world's most prominent immigration country of all time.[102] This open immigration was a major driver of American innovation, industry and growth.

Fruits of Open Immigration—Innovation All Around

The United States of the nineteenth century was still a rural society for the most part. One of the unique features of the Industrial Revolution in the United States that differed from that in Britain was the geographical dispersal of industrialization throughout the US countryside.[103] During the early nineteenth century, "rural" in the United States became synonymous with industrialization rather than with traditional agriculture.[104] Industrialization also became synonymous with internal migration.

After the American War of Independence in 1776, industry-oriented farmers in New England and the Middle Atlantic sought to gain wealth through trade, even if it meant in unfamiliar places with unfamiliar people. They started making long trips in search of new markets for their products and crops.[105] During the first half of the nineteenth century, more sellers, distributers and buyers across the country learned that the key to opportunity and wealth might await them far from home. By the 1820s, long-distance marketing trips gave rise to a network of entrepreneurial farmers and peddlers that dispersed around the countryside.[106]

The emerging footwear industry in New England serves as a good example for the effect of open migration on the rise of an industry. Lynn, a small town near Boston, was traditionally home to a small community of shoemakers. After the American War of Independence, a local entrepreneur started supplying shoes to slaves in Southern states, without regard to the ideological and cultural gap between the American North and South. In this manner, New England-made standardized and inexpensive shoes represented some 60 percent of all manufactured shoes across the entire country by 1860, against the backdrop of the Civil War.[107]

The railroad industry, especially in the North and as it expanded westward, served as an efficient vehicle for innovation. First, the booming railroad industry employed machinists from other machinery sectors, mostly the textile industry. In this way, the railroad industry—coupled with the domestic open attitude—enabled versatility among skilled workers of all social backgrounds, who moved from one job to the other across US industrial centers. After 1835, three-quarters of innovators with textile-related training moved to and innovated in other sectors, occasionally far from their place of origin.[108] Innovation and industrialization developed less in the Southern states because of the captive labor force of slaves.[109] Besides the inherent evil of the system, slavery—with its cheap labor costs—provided a disincentive for mechanization, industrialization and innovation.

Based on their openness to travel, people and ideas, innovators in the United States passed on technological knowledge, not only from one sector to another but also from one region to another. For instance, three-quarters of Ford's labor force comprised foreign-born workers.[110] These immigrant job seekers, along with their American-born counterparts, traveled across the country to Detroit to become Ford workers. They all filled Detroit with a labor force that other automobile industries would employ later on.[111]

In the late eighteenth and the nineteenth century, the immigration of new peoples and their easy migration within the United States fueled demand for all sorts of products. At the same time, it created a growing labor pool that could meet that demand. Most importantly, it created a mix of peoples, cultures and backgrounds that provided a fertile ground for creativity, innovation and entrepreneurship. This was not the situation in Mexico and Brazil.

Mexico: Selective Immigration Stymies Innovation and Growth

Mexico, a sparsely populated country like the United States, would also have greatly benefited from immigration. Nonetheless, Mexico evolved differently. Mexico's closed attitude toward immigration ensued from its monopolistic, nonempowering economic system and enduring traditional culture. For several decades after Mexican independence, immigration to Mexico remained limited.

First, the largely Catholic Mexican society was xenophobic toward and discriminated against Protestants. The Mexican church, which had grown extremely wealthy during the colonial era, further reinforced Mexico's opposition to multicultural immigration. Following Mexico's independence in 1821, clergymen across the country invested much energy to earn incomes from their religious status: priests owned farms, ranches, expensive herds of cattle and horses, and country stores.[112] While monopolizing access to lands, Mexico's religious elite restricted access to those not closely connected to the Catholic clergy. This exclusion was even stronger for non-Catholic immigrants and investors from overseas.

The best lands in Mexico remained in the hands of several local economic elites, who were not welcoming toward newcomers who potentially threatened their traditional status; such newcomers represented a competitive agricultural labor force to the long exploited local indigenous populations.[113] Thus, Mexico remained unattractive to immigrants from Europe and elsewhere fleeing religious persecution and poverty. Such immigrants would avoid the unfair competition for arable land and seek opportunities elsewhere, especially the United States.[114] It is not surprising then, that during the nineteenth century, the Mexican population increased only from five to fifteen million, while the US population increased from five to ninety million.[115]

Moreover, migration within Mexico was limited, because the Mexican political structure restricted large investments in transportation infrastructure. As seen earlier, in the United States, entrepreneurial business could rather easily invest in such infrastructure with the support of the US government. Until the construction of an efficient railroad network in Mexico during the late nineteenth century, the country's interior was practically landlocked from the centers of industrialization in Europe and North America. The cost of transportation from London to Mexico City was much more expensive than from London to New York and other industrial centers in North America that were networked by efficient canals and waterways.[116]

For these reasons, by the end of the nineteenth century, only 0.6 percent of all European immigrants in the world chose to immigrate to Mexico.[117] Relative to the United States, Mexico was not very open to immigrants, as it had selective immigration, limited migration and limited opportunities. In light of numerous constraints, the country's elite immigrants retained their native languages and even saw their native culture as superior to that of the masses in their host country. Immigrants to Mexico avoided intermarriage with local Mexicans and maintained segregated communities. Moreover, only a few hundred immigrants applied for Mexican citizenship. The rest relied on diplomatic protection.

Mexico therefore did not really become a "melting pot" for masses of the world's immigrants like the United States, but rather lacked the openness to peoples, perspectives and ideas so essential for innovation. In Mexico during the late nineteenth century, only few European (as well as American) immigrants started to engage in entrepreneurial-oriented immigration.[118] European entrepreneurs avoided starting new families or calling up their families from their homeland to resettle in Mexico. They were mostly young men in their twenties who stayed in Mexico for a limited time. In fact, 75 percent of all immigrants in Mexico at the time were men.

The Mexican government, mostly during the rule of General Porfirio Díaz (1876–1911), attempted to attract European immigrants to populate its vast land. However, its immigration goals fell short because the country was not open to all types of populations. Mexico's government was also not willing to empower newcomers. It mostly sought the already empowered individuals who could contribute to its fledging economy. With the exception of General Díaz's rule, the Mexican government did not launch large development projects that granted lower-class Mexicans or immigrants a share of its abundant land in the North.[119]

As a result, a few wealthy French and Spanish immigrants' families monopolized the industrial efforts in the country. Some European immigrants seized control of the developing banking sector and commerce.[120] These immigrants were wealthy investors rather than poor and dispossessed workers. Just the reverse occurred in the United States, where masses of laborers came with dreams of wealth and flourished through hard work and entrepreneurship. Ironically, the US system that practiced open immigration empowered all immigrants and encouraged competition to produce an environment where millions of poor immigrated, worked hard, started businesses and created wealth. In Mexico, selective immigration restricted opportunities for advancement, and restricted competition created a relatively fossilized society with much less innovation, entrepreneurship and economic progress.

Mexico witnessed its immigration peak only in 1930, but the Mexican government gradually closed the doors to immigration during the twentieth century.[121] Immigrants made up less than 1 percent of the country's population in 1930. Further, due to a lack of opportunity, twentieth-century Mexico witnessed the turning of the nation into a net supplier of emigrants, most of whom went to the United States. In doing so, much of the Mexican labor force enriched other economies rather than their own economy.

After the Mexican-American War, the United States annexed several Western states, including Texas and California. It provided these new states the same rights as the original states of the union. In particular, the states enjoyed the same level of openness, empowerment and competition that the rest of the United States enjoyed. After becoming a part of the US union, these states—especially California—became the world's most welcoming regions of mass immigration and hubs of industrial and agricultural innovation.[122] The new Western states witnessed an influx of immigrants, a tremendous boost in their economies and a flourish of innovation. The incorporation of states into the United States reflects a great natural experiment, contrasting the policies of Mexico and the United States. Under Mexico, the states were relatively dormant. Under the United States, however, they grew to become major economies in their own right.

Brazil: Selective Immigration Despite Sparse Population

Development in Brazil was comparable to that in Mexico. In Brazil during the early nineteenth century, large plantation owners imported African slaves to cultivate their lands.[123] Between 1800 and 1852—despite increasing pressure from the British Empire on Brazil to stop slavery—landowners coerced around 1.3 million slaves to work in the country.[124] Thus, wealthy plantation owners, well connected to the political elite, did not fill their need for labor through the migration of domestic free workers or the immigration of overseas free workers.

Under such circumstances few free laborers were willing to immigrate to Brazil's frontiers, where slavery was the norm.[125] In addition, as seen above, the nonempowering Land Law of 1850 restricted—both de facto and de jure—the allure of the country's vast unpopulated lands in the eyes of immigrants. The law practically denied landless farmers, who did not belong to the landed elite, easy access to farms—not to mention their right to own these lands. Newcomers could only come as low-wage workers or slaves, and compete with the local labor force.[126]

With such limiting conditions, Brazil suffered from a critical shortage of skilled labor for industry in the first half of the nineteenth century. Brazil received fewer than 50,000 immigrants during the first half of the nineteenth century, at the time when the United States had 35 million immigrants.[127]

As in Mexico, another major factor that hampered Brazil's willingness to receive laborers from overseas was its closed attitude toward diverse populations of different ethnic and religious backgrounds. Unlike the United States, where state and church were constitutionally separated, the Brazilian Constitution of 1824 identified Catholicism as the religion of the state. It denied non-Catholics the right to acquire higher education and employment in governmental positions—whether executive, legislative or judicial. Moreover, a royal decree forbade non-Catholics from building or attending worship houses, such as non-Catholic churches, temples or synagogues.

Within this restrictive atmosphere, numerous non-Catholic immigrants, as well as Jews from Europe, avoided immigrating to Brazil. Thus, between 1820 and 1890—a year after the founding of the First Republic of Brazil when the government implemented this discrimination policy—the number of immigrants settling in Brazil was relatively small: only 876,980. Of this number, some 764,687 (87 percent) came from Portugal, Spain and Italy, countries in which the Catholic faith was prominent.[128] Moreover, Brazil's declared intention throughout the second half of the nineteenth century to "whiten" its population further hampered other immigrant aspirations, such as Asian immigrants.[129]

Only by the late nineteenth century, following the 1889 Revolution when slavery was truly abolished in the country, did Brazil open its immigration policy to attract new immigrants to its coffee plantations. After that, Brazil became the second largest immigrant-receiving country in Latin America, after Argentina. Out of the 11 million immigrants who came to Latin America between 1851 and 1924, 33 percent entered Brazil, while only 3 percent chose Mexico, which still had its closed-door policy.[130]

Thus, between the seventeenth and the nineteenth century, Brazil remained a closed country with vast lands and opportunities, but limited openness, empowerment and competition, which consequently limited immigration, industry and innovation. The burst in immigration during the last decade of the twentieth century enriched the population of Brazil and triggered renewed industrialization. Openness to immigration was one of the factors that turned Brazil into the most economically successful nation in South America. However, this shift seems to have appeared too late for Brazil to keep up with the United States. By the late nineteenth century, the United States quickly surpassed Brazil's immigrant population, becoming the biggest immigrant-absorbing country in the world and in modern history. The US open policy, combined with the competition and empowerment of peoples, led the United States to have a dominant economy relative to Brazil, though both started on an equal footing at the beginning of the nineteenth century.

Decline of the United States?

The United States seems to have declined in relative economic power from its peak in the decade following World War II (see Figure 11.3). This decline can partly be attributed to the relative

Figure 11.3 Share of world GDP from 1700 to 2008

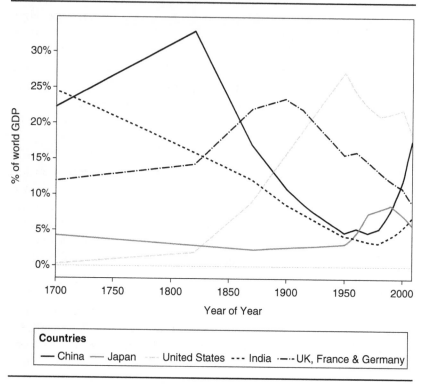

Source: https://infogr.am/Share-of-world-GDP-throughout-history. M Tracy Hunter/Wikimedia Commons.

rise of Western Europe in the 1950s, Japan in the 1970s, Southeast Asia in the 1980s, China in the 1990s and 2000s, and India in the last decade. Indeed, by 2017, China became a larger economy in GDP by purchasing power parity (PPP) than the United States, while India ranked third.[131] The rise of all these nations relative to the United States reflects the diffusion of technology and appreciation for the driving forces of openness, competition and empowerment to less developed nations. In particular, Japan, Southeast Asia, China and India experienced rapid growth when they changed from relatively closed economies to open, highly competitive markets. As a result, in terms of relative GDP, the United States is less dominant now than it was immediately after World War II.

So, is the United States really in decline? To understand this issue, one needs to examine the current state of the country in terms of the very same three forces that led to its rise: openness, competition and empowerment.

First, consider openness. The United States continues to be the beacon for innovators, refugees and immigrants from all over the world. As of 2015, the United Nations estimates that 47 million people living in the United States were not born there.[132] This number is four times that of the next largest immigrant population in Germany.[133] In fact, about one-sixth of American residents are immigrants. Thus, until January 2017, when a new administration assumed power in the United States with distinctly conservative views on immigration, the United States continued to be the dominant destination country for immigrants. Even in 2017, its position as a prime destination for immigrants in the world is unlikely to change much, given the resistance of many groups and courts to President Donald Trump's feeble executive orders restricting immigration. Two other phenomena support the role of immigration in entrepreneurship and innovation.

One, immigrants have made a mark in the United States as entrepreneurs and innovative business owners, at much higher rates than have people born in the United States.[134] For example, 40 percent of the companies on the Fortune 500 list have been founded by immigrants or children of immigrants.[135] Similarly, immigrants have founded almost half the billion-dollar start-ups in the United States.[136] This criterion of billion-dollar start-ups is important because it indicates the start of new businesses that disrupt incumbents and transform markets. The high proportion of immigrant founders in these start-ups is quite impressive, given that most immigrants came to the United States either poor, with very little capital or after having suffered severe trauma in their country of origin. Though immigrants consist of only 15 percent of the US population, they account for 40 percent to 50 percent of start-ups, depending on the measure used.

Two, Silicon Valley in the United States is a region of vibrant innovation and entrepreneurship.[137] About 52 percent of the start-ups in Silicon Valley between 1995 and 2005 were founded by immigrants.[138] Indeed, regarding immigration and entrepreneurship, the United States and Silicon Valley in particular represent a natural experiment. Our research suggests that the proportion of immigrants who start successful businesses in the United States is higher than the proportion of those in the country they left behind. Why so? Part of the reason is that immigrants—by the very act of immigration—are selectively more entrepreneurial, while the changes in environment and new networking opportunities stimulate further innovation and entrepreneurship.[139] However, we posit that the primary reason is that the level of competitiveness and empowerment in the United States is much higher than that in the countries the immigrants left behind. We address these two issues next.

Second, consider competition. The United States remains one of the most competitive nations in the world. In the 2015–16 rankings of competitiveness, the United States ranked third out of 140 countries and first among large countries.[140] The only two countries that ranked above the United States were Singapore and Switzerland, both tiny in comparison. Two large rapidly growing nations ranked much lower: China ranked twenty-eighth and India ranked fifty-fifth. One empirical indicator of US competitiveness is its trade surpluses. Indeed, many major economies in the world run major trade surpluses with the United States. One cause of these surpluses is that the United States is relatively open to imports. Indeed, the nation may be more open to imports from certain countries than those countries are open to exports from the United States. Primary cases of these are China, Japan and Mexico. Even Europe as a whole and Germany, in particular, run trade surpluses with the United States. This was one of the major issues in the 2017 presidential election campaign in the United States, which Trump won by a small margin.

Another indicator of competitiveness is that products are available in the United States at prices lower than those in home nations of the exporters.[141] This applies to Chinese, Taiwanese, Japanese, South Korean, Israeli and various European imports. In effect, this price asymmetry means that either companies in foreign firms are charging US consumers less than what they are charging consumers in their home markets, or that foreign nations are levying various taxes and tariffs on the sale of goods in home markets more than what the US government levies on those same goods domestically. While this situation is unfortunate for US firms and US workers, it stimulates innovation, as firms in the United States strive to produce goods that are more innovative and have higher value added than those produced by foreign firms.

For example, production of low-tech goods, such as textiles, clothing, shoes, furniture, small tools and so on may have vanished in the United States. Production of even some moderately high-tech products, such as solar panels, steel, automobiles, home appliances, televisions and so forth has left or may be leaving the United States soon. However, the United States is still very dominant in the production of high-tech products such as spacecraft, aircraft, advanced defense equipment, self-driving cars, auto batteries and so on. Similarly, while the production of cell phones may have moved abroad, their design still takes place in the United States. Most importantly, most of the major companies responsible for the digital revolution originated in the United States. For example, Microsoft, Intel, HP, Dell, Google, Amazon, Facebook, Apple, Uber and Airbnb, all started in the United States. Further, a study of billion-dollar start-ups shows that two-thirds of these occur in the United States (see Figure 11.4). Start-ups are important because they reflect innovativeness. Billion-dollar start-ups are important because they indicate innovation that has already won market acceptance as reflected in its large valuation.

Third, consider empowerment. Even in the turbulent economic environment of the 2010s, the United States remains a friendly place for immigrants to study, progress, start businesses and advance in careers. American universities have an explicit policy of not discriminating by country of origin. Thus, people from all over the world can come to the United States and get an education at top American universities, in competition with domestic students. Once they graduate, employee-friendly laws enable them to get high-paying jobs that are superior to those they would get in their home countries, even with an American education. These immigrants are motivated to stay in the United States and contribute to its growth, just as immigrants did centuries before. Further, most companies are open to recruiting international students just as they might domestic students. This

Figure 11.4 Billion-dollar start-ups by country of registration

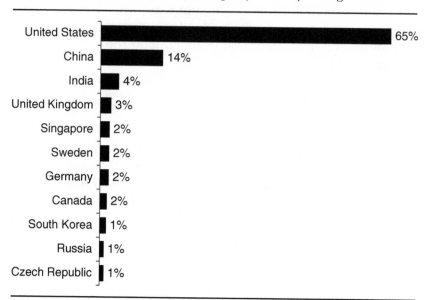

Source: Gerard J. Tellis, 2016, "Startup Index of Nations," Working Paper, USC Marshall School of Business.

openness is at least partly motivated by the fact that many of these companies are international and want an international workforce.

In addition, banks, venture capital firms and other investors are comfortable providing immigrants with capital to start businesses. When banks and investors are unwilling to do so, communities or prior established immigrants from a country form networks that loan to new immigrants from that same community.[142] Thus, established immigrant networks foster entrepreneurship and innovation in the new immigrant population. All the above factors establish an environment in which the United States continues to be an exceptionally open, empowering and yet competitive melting pot of immigrants with different ethnicities, cultures, religions and national origins. This mixing itself strengthens the fabric of society and stimulates innovation.

In conclusion, threats surely loom on the horizon to the dominance of the United States in terms of global GDP and innovation: a backlash against immigrants among some disadvantaged segments of society in the United States; a new administration with conservative—if not anti-immigrant—policies; slow economic growth in the United States; and fast economic growth in India, China and many other rapidly emerging markets. Yet, as of now, the three forces of openness, competition and empowerment are still strong in the United States, promising a fertile climate for innovation and entrepreneurship. This analysis suggests obvious implications for the United States in terms of openness to immigration, empowerment of domestic and international peoples, and competition among start-ups and established companies.

12

Lessons

The previous chapters are an attempt to portray the nature of transformative innovation. The premise we offer is not one of particular events in specific societies or regions. Rather, we suggest that focusing on specific events, which are particular to one period or region, could miss the bigger—much richer—picture. We offer a perspective that is universal in nature: Whereas each transformative innovation was developed in a different period, region and social and economic context, they all share basic traits. Namely, they all propelled the societies that developed and embraced them to meaningful dominance, and they all relied on varying degrees of openness, competition and empowerment.

This survey of world history over the last 2,000 plus years reveals important patterns that provide lessons for modern leaders and policy makers.

1. *Irresistibility of Innovation. No single nation can stop the path of innovation.* Any single nation may decide not to embrace innovation and new technologies. However, the world is full of diverse countries, cultures and peoples. Those not at the pinnacle of power and wealth may have a strong desire to get there. This aspiration may make nations eager to embrace or develop innovations that can help their advance. Thus, even if a dominant nation ignores or suppresses innovation, another nation or region will out-innovate it. One can see this pattern most easily in the case of gunpowder weapons

and water navigation. Supremacy in gunpowder weapons shifted between the Jin, Song, Mongol and Ming dynasties in China, then moved to the Ottomans in Turkey and to a number of nations in Europe (see Chapter 4). Each nation that innovated gained an edge in competitive struggles. But because the technology had low entry barriers, other nations were quick to imitate and further innovate. The Japanese samurai and Egyptian Mamluks, who tried to stop gunpowder weapons, failed dramatically. A similar pattern occurred with water navigation. Here, supremacy moved from China, to Venice, to Portugal, to the Netherlands and to England (see Chapters 5 to 8).

2. *Transformation from Innovation. Certain innovations change the economics of cities, regions, nations and the world, sometimes transforming small cities, regions or states into world powers.* For example, ancient Rome was one of many city-states around the Mediterranean that was able to rise to power. Roman mastery of concrete enabled the building of infrastructure and communication for and administration and cultural cohesion of a great empire that dominated the countries around the Mediterranean Sea—from England to the Middle East (see Chapter 2). In the thirteenth century, Mongolia was a remote and seemingly resource-poor region of Asia. The development of swift equine warfare enabled Genghis Khan to transform the Mongolian horse into a powerful equine fighting machine and build a great empire (see Chapter 3). Similarly, prior to the innovations of the patent system and the steam engine, the island nation of England was one of several sparring nations in Europe. However, the embrace of these innovations ahead of other countries of the world, transformed England into a powerful nation and enabled it to build a global empire (see Chapters 9 and 10).

3. *Transience of National Dominance. No nation's dominance in the world is permanent.* Rather, small city-sates, regions and nations rose to become great empires and then fell back to a low status or were engulfed by newly emerging nations and empires. Chapters 2 to 10 traced the transient dominance of Rome, Mongolia, China, Venice, Portugal, the Netherlands

and England (For a graphic, see Figure 11.3 in Chapter 11 and Figure 9.1 in Chapter 9). While the United States is currently dominant, threats to its hegemony loom on the horizon (see Chapter 11). While this pattern may seem natural, it runs counter to the innovation-wealth-innovation hypothesis.[1] Innovations lead to improvements in human living standards, efficiencies in production of goods and services, advantages in international trade and ultimately great wealth. Such wealth provides the innovative region or nation with the means to attract talent, support infrastructure and fund research for still further innovation (see Chapter 1). Due to this positive feedback loop, a once successful or dominant nation should remain permanently dominant. Indeed, when one country is dominant, its power and position seem invulnerable. However, decline and the rise of other nations invariably seem to arrive. Why? The main reason for this pattern is the rise of transformative innovations in rival nations due to the increase in openness, competition and empowerment (see Chapters 4 to 11).

4. *Disruption from Obstruction. A dominant nation that ignores or obstructs innovation falls behind and is disrupted, despite its wealth and power.* Typically, this situation occurs due the rise of a new transformative innovation in some other—likely competing—region or nation. Once gunpowder became standard, the advantage of equine warfare greatly diminished (see Chapter 4). Warriors could stop the advance of horses from afar. Nations that embraced gunpowder sustained or advanced their status. Those that ignored it succumbed to powers that did embrace it. The Mongols succeeded in China (as the Yuan dynasty) only because they embraced and advanced gunpowder weapons. Likewise, various innovations transformed the fortunes of Venice, Portugal, the Netherlands and England. The Venetian galley and its unique Arsenal, which manufactured galleys at a rate of about one a day, made Venice the dominant force in the Mediterranean (see Chapter 6). The galley also enabled Venice to control the trade in spices and silks from the East to Europe. However, the Portuguese caravel enabled a direct sea route to India and the East, cutting off the middlemen (Arabs,

Turks, Mamluks and Venetians), and brought spices and silks directly to Europe at competitive prices. The transformative innovation of the Portuguese caravel rendered the Venetian galley obsolete (see Chapter 7). Venice eventually became a museum.[2] Similarly, the Dutch *fluyt* rendered the Portuguese caravel obsolete and helped the Dutch transform the Netherlands into the premier nation in world trade (see Chapter 8). Likewise, the steam engine, especially the British steamship, rendered the Dutch *fluyt* obsolete and helped the British become the next dominant sea power and trading nation (see Chapter 10).

5. *Curse of Success. Success sows the seeds of failure.* Transformative innovations and the ensuing dominance generate enormous wealth and success. However, success prompts the innovative nation to focus on advancing the one technology that made it successful and to the negligence of other emerging technologies. Success blinds the successful nation to the technological horizon, where future innovations emerge that could render the current technology obsolete. Typical examples are Venice, Portugal and the Netherlands—all countries that rose on the wave of a transformative innovation, but then failed to foresee, embrace and capitalize on the next innovation in this sequence. For example, Venice developed the Arsenal and the galley that it enabled it to dominate trade in the Mediterranean with Asia through Africa and the Middle East (see Chapter 6). But it failed to develop anything like the caravel that enabled circumnavigation of Africa and trade in the Indian Ocean directly with Asia. Portugal developed the caravel that enabled direct trade with Asia through the Atlantic and obsoleted the Venetian galley (see Chapter 7). The Dutch developed the *fluyt*, a more efficient ship for trade in the oceans, which obsoleted the Portuguese caravel (see Chapter 8). In some cases, preoccupation with perfecting the current technology that triggered the nation's rise blinded it to a new technology that obsoleted the current one.

A similar pattern exists in the modern age at the corporate level. Small start-ups rise on the great success that flows from an innovation. However, once successful, the corporation is so focused on perfecting the past innovation that brought it success, that it is blind to future innovations that might obsolete it if ignored. Ironically, some of these new innovations originated deep in the labs of these very companies. Examples abound.[3] Xerox developed most of the innovations of the computer age but was so focused on the technology of copying that it did not commercialize computer innovations. Thus, it missed out on personal computers and printers that led to the rise of Microsoft, Intel, HP and Dell. Kodak held most of the patents in digital photography. But it was so focused on analog (film) photography that it failed to successfully commercialize digital photography. The latter became pervasive through cell phones, leading to the rise of Apple and Samsung. HP was so focused on printers and PCs that it failed to see the importance of tablets and mobile phones for copying, sharing and reading. Nokia and Blackberry pioneered mobile phones and smartphones respectively but failed to see the importance of the touch screen smartphone commercialized by Apple. The cycle of the rise and fall of corporations based on a succession of disruptive technologies is seemingly endless.[4]

Two deep organization traits and practices account for the curse of success: organizational filters, routines and investments.[5] These have been studied primarily at the corporate level but would likely apply at the national level. One can think of the nation as a larger entity than the corporation that suffers these limitations analogously. Organizational filters are procedures for receiving, screening and interpreting information signals from the outside world. These filters are technology specific. So, when the current technology is the galley, all information received is processed to advance the galley rather than to obsolete the galley. As such, information signals that may obsolete the galley would not be read or may be misinterpreted. Thus, the focus on the galley lives on. Organization routines are practices that perpetuate the success of a current technology. Venice's Arsenal had numerous practices for the selection, training and advancement of workers for the successful functioning of the galley (see Chapter 6). Such routines would make the identification and development of a rival technology very unlikely. Organization investment is the resources sunk into developing a past innovation. Such investments are, in effect, sunk costs that cannot be recovered if the passed innovation is no longer relevant. Venice's Arsenal is a prime example of an organization investment. Venice had invested in and perfected the Arsenal to such an extent that at its peak it could turn out about one galley a day. With such a huge investment, the nation was less inclined to pay attention to a new technology such as the caravel that may not be produced easily in the Arsenal. Thus, organization filters, routines and investments can perpetuate the success in one technology at the cost of identifying and developing a rival, potentially superior technology.

6. *Transience of Resources. Geography and natural resources provide a passing but not perennial advantage dependent on technological innovation.* Geography and natural resources were located in regions that rose to power years before their rise. The arrival of a critical innovation transformed a potential opportunity, in some cases a geographic one, into an active advantage. Prior to the innovation, resources may have seemed natural, ordinary or even a waste. After the development or adoption of the innovation, however, what was once waste became valuable. For example, prior to Genghis Kahn, the Mongolian horse was primarily a wild resource tamed for local food and transportation. After the development of swift equine warfare, it became a powerful resource for building the empire (see Chapter 3). After the development of gunpowder and later the steam engine, the value of the horse greatly diminished (see Chapter 4). Similarly, prior to the arrival of mechanization and the steam engine, England's iron ore and coal reserves had limited use. The development of the steam engine helped propel the nation to world leadership (see Chapter 10). However, today with the development of cleaner fuels such as petroleum, nuclear, wind and solar, coal resources are again no longer critical for a nation's success.

The perceived importance of resources arises from a static perspective that fails to consider dynamics. For example, today, the oil states of the Persian Gulf seem resource rich. But when viewed through the lens of time (considering dynamics) their wealth seems passing. Before the arrival of mining technology for oil, many of these states seemed resource poor. With the arrival of fracking technology in the United States, oil prices have dropped and so have the seeming power and wealth of these states. As the technology for solar and wind energy develops further and their prices drop, the oil states will seem less wealthy. So geography and natural resources create wealth primarily by the availability of relevant technological innovation. Superior innovation can devalue that wealth as quickly as some innovation rendered it valuable.

7. *Limits of Religion, Geography and Luck as Causes. Prior explanations for the rise to dominance of nations, such as religion, geography or luck, crumble when one considers 2,000 years' dominance of various nations.* Such prior explanations include culture, religion, geography, climate, colonization and luck (see Chapter 1). During 2,000 years, both non-Christian peoples (Romans, Mongols, Chinese and Ottomans) and Christian states (Venice, Portugal, the Netherlands and England) rose to be powerful global empires. Likewise, such powers have been Catholic (Venice, Portugal), non-Catholic (the Netherlands, England) and avowedly nonreligious (United States). No major climate or geographic changes occurred during the rise or fall of these major nations. As documented in the prior chapters, the only common, major event that occurred concurrently with the rise in the dominance of these major nations is the rise of a transformative innovation. In most cases, a major factor that triggered their decline was another transformative innovation adopted by or developed in competing nations. Alternatively, the three drivers that facilitated innovation in the focal nation declined and gave way to their ascendance in a competing region. Thus, change in innovation seems to be the most important common explanation for the rise in dominance of nations.

The diffusion of innovations is evident from very early periods. This fact diminishes the role of geography because from early on, it is evident that the specific region of origin of a transformative innovation is not very important. Far more important is what nation or region adopted the innovation and continued to develop it, master its technology and put it into effective use, thereby facilitating further innovation.

The argument of luck is subtler to refute. Luck implies being "at the right time and place." To paraphrase, it implies that a nation rose because of a fortuitous combination of events and locations. But therein lies its fallacy. The argument of luck is essentially circular. It identifies no exogenous factors of causation. It merely labels the combination of time and place as fortuitous because those events had beneficial outcomes. However, our tracing of the rise of transformative innovations shows that they occurred over a long time—sometimes over hundreds of years—during which they built up slowly. Visionary leaders sometimes charted their course. At other times, innovators endured great hardships and risks to achieve the end result. There was no specific, swift fortuitous moment and place where innovative success suddenly occurred. Innovation is neither random nor serendipitous. Instead, it is the fruit of a conducive environment[6] characterized by three institutional drivers: openness, empowerment and competition.

8. *Preeminence of Institutional Drivers. Across 2,000 years, three institutional drivers were prevalent to varying degrees when transformative innovations took off, and absent when they did not: openness, empowerment and competition.* We have charted the role of these three factors in the rise of all of the ten histories. In some cases, the evidence for all three drivers is rich, while in other cases, the evidence on a specific driver is sparse. In the history of the rise to dominance of the United States, the role of these three factors is most clear because of the existence of two counterexamples: Brazil and Mexico. In comparing the United States, Brazil and Mexico, we have a natural experiment (see Chapter 11).

Around 1776, when the United States became a nation, Brazil and Mexico were larger and more promising economies than the United States. Both had larger populations than the United States. Both were superior to the United States in terms of known resources at the time, having more easily accessible gold or silver, and more arable land for cultivation of sugar and cotton. Within a span of 150 years, the United States sped ahead to become a world superpower. Mexico and Brazil also grew but much more slowly. Chapter 11 argues that the United States adopted mass production, an innovative new way to manufacture, distribute and market products and services. Key drivers of this transformative innovation were openness, empowerment and competition. The crux of the argument is as follows: while Mexico and Brazil were open to immigrants of only one nation or one region, the United States was open to most peoples, initially of Europe and later of the whole world. While Mexico and Brazil provided land to the elites while disempowering others, the United States empowered most of its citizens to own and cultivate land, or to start other enterprises. And while Mexico and Brazil restricted competition by protecting the monopoly of elites, the United States fostered competition and explicitly controlled monopolies. Intense competition ensued. This policy of openness, empowerment and competition triggered immense innovation and entrepreneurship, especially in standardization and mass production, which permeated every industry. A modern, dynamic mass-market economy was born, becoming the template for nations all over the world.

Several recent authors have predicted the decline of the United States and the rise of other nations, especially China and India.[7] Interestingly, the economics of these two countries took off when they embraced competitive markets: China in 1978 and India in 1992. However, we posit that the future path of a country should not be predicted based on past or even current trends. If so, the only prediction is that the leading country may be surpassed by another. Rather, the future path should be predicted by which country will best embrace openness, empowerment and competition. At present, the United States still seems to have an edge (see Chapter 11). But the present is rapidly changing. The future is in the hands of today's leaders and policy makers.

Notes

1 Global Influence of Transformative Innovation

1 The ships had multiple watertight compartments, movable sails, a sternpost rudder and compass. The Chinese had also other innovations in gunpowder, cannons, agriculture, ceramic and silk.

2 Maddison Project Database, version 2013. Jutta Bolt and Jan Luiten van Zanden, "The Maddison Project: collaborative research on historical national accounts," *Economic History Review* 67, no. 3 (2014): 627–51.

3 Stanley L. Engerman and Kenneth L. Sokoloff, "Factor Endowments, Institutions, and Differential Paths of Growth among New World Economies," in *How Latin America Fell Behind: Essays on the Economic Histories of Brazil and Mexico, 1800–1914*, ed. Stephen H. Haber (Stanford, CA: Stanford University Press, 1997), 260–304.

4 For example, see Steven Johnson, *Where Good Ideas Come from: The Natural History of Innovation* (New York: Riverhead Books, 2010).

5 Alejandro Portes, "Economic Sociology and the Sociology of Immigration: A Conceptual Overview," in *The Economic Sociology of Immigration: Essays on Networks, Ethnicity, and Entrepreneurship*, ed. Alejandro Portes (New York: Russell Sage Foundation; Rosenzweig, 1995), 1–41; Stav Rosenzweig, Amir Grinstein and Elie Ofek, "Social Network Utilization and the Impact of Academic Research in Marketing," *International Journal of Research in Marketing* 33, no. 4 (2016): 818–39; Adrian Furnham, "Why Immigrants Make Great Entrepreneurs," *Wall Street Journal*, November 26, 2017. Gerard J. Tellis and Stav Rosenzweig, "On Immigration, Do as the Romans Did," *Wall Street Journal*, February 7, 2018.

6 Wendy Duivenvoorde, *Dutch East India Company Shipbuilding: The Archaeological Study of Batavia and Other Seventeenth-Century VOC Ships*, (College Station: Texas A&M University Press, 2015).

7 Jared Diamond, *Guns, Germs, and Steel: The Fates of Human Societies* (New York: W. W. Norton, 1999).

8 For example, see Philip M. Parker, *Physioeconomics: The Basis for Long-Run Economic Growth* (Cambridge, MA: MIT Press, 2000); Philip M. Parker, "Climatic Effects on Individual, Social, and Economic Behavior," *Physioeconomics Review of Research Across Disciplines* (Westport, CT: Greenwood, 1995).

9 Gerard J. Tellis, Stefan Stremersch and Eden Yin, "The International Takeoff of New Products: Economics, Culture and Country Innovativeness," *Marketing Science* 22, no. 2 (Spring 2003): 161–87.

10 The most influential work is the 1905 classic study by Max Weber: *The Protestant Ethic and the Spirit of Capitalism.* The following represent works that are more recent. For a strong argument in favor of the dominance of the "West," see Ian Morris, *Why the West Rules—For Now* (New York: Farrar, Straus and Giroux, 2010). For a treatise about the role of the Christianity and the West see, Rodney Stark, *How the West Won: The Neglected Story of the Triumph of Modernity* (Wilmington: Inter-Collegiate Institute, 2014); Rodney Stark, *The Victory of Reasons: How Christianity Led to Freedom, Capitalism, and Western Success* (New York: Random House, 2007); David S. Landes, *The Wealth and Poverty of Nations: Why Some Are So Rich and Some So Poor* (New York: W. W. Norton, 1998). For a treatise about the role of Western culture, see Joel Mokyr, *A Culture of Growth: The Origins of the Modern Economy* (Princeton, NJ: Princeton University Press, 2016).

11 In discussing the triumph of the West, Ian Morris uses a very liberal definition of the West that includes Western Europe, North America, ancient Greece and Rome, and even Egypt, Mesopotamia and Ottoman Turkey. Even with such a broad definition, the theory still fails to explain the wealth of China and India until 1500 (and as some scholars, such as Gunder Frank, would suggest—until 1800) and the recent rise of China.

12 Some authors argue that the dominance of China and India continued well into the seventeenth century. See A. Gunder Frank, "Central Asia's Continuing Role in the World Economy to 1800," in *Historical Themes and Current Change in Central and Inner Asia*, ed. M. Gervers and W. Schlepp (Toronto: University of Toronto, 1998): 14–38.

13 Mokyr, *A Culture of Growth.*

14 Kenneth Pomeranz, *The Great Divergence: China, Europe, and the Making of Modern World Economy* (Princeton, NJ: Princeton University Press, 2000).

15 See, for example, Sucheta Mazumdar, "The Great Divergence: China, Europe, and the Making of Modern World Economy—Review," *Technology and Culture* 44, no. 3 (2003): 604–6.

16 Daron Acemoglu and James Robinson, *Why Nations Fail: The Origins of Power, Prosperity, and Poverty* (New York: Crown Books, 2012).

17 The right time and place argument is circular and nonfalsifiable. Hence it has no explanatory value. Consider the combination of time and place has to be either right or wrong. However, it cannot be "wrong," because the innovation was adopted. Therefore, it has to be "right." Hence, "right time and place" is merely after the fact naming of the time and place of the adoption of the innovation. It is true by definition. So it is circular. It cannot

be false. Therefore it is non-falsifiable. The key question is what makes it the "right time and place?"

18 Richard Wiseman, "The Luck Factor," *Skeptical Inquirer*, May/June, 2003.

19 Gregory J. Feist, "A Meta-Analysis of Personality in Scientific and Artistic Creativity," *Personality and Social Psychology Review* 2, no. 4 (1998): 290–309.

20 Jeffrey Dyer, Clayton M. Christensen and Half B. Gregersen, *The Innovators DNA: Mastering the Five Skills of Disruptive Innovators* (Cambridge, MA: Harvard Business Review Press, 2011); Abbie Griffin, Raymond L. Price and Bruce Vojak, *Serial Innovators: How Individuals Great and Deliver Breakthrough Innovations in Mature Firms* (Palo Alto, CA: Stanford Business Books, 2012); Gerard J. Tellis and Peter N. Golder, *Will and Vision: How Latecomer Grow to Dominate Markets* (New York: McGraw Hill, 2003).

21 Stav Rosenzweig and David Mazursky, "Constraints of Internally and Externally Derived Knowledge and the Innovativeness of Technological Output: The Case of the United States," *Journal of Product Innovation Management* 31, no. 2 (2014): 231–46; Stav Rosenzweig, "Non-Customers as Initiators of Radical Innovation," *Industrial Marketing Management* 66, no. 6 (2017): 1–12; Stav Rosenzweig, "The Effects of Diversified Technology and Country Knowledge on the Impact of Technological Innovation," *Journal of Technology Transfer* 42, no. 3 (2017): 564–84.

22 Gerard J. Tellis and Peter Golder, *Will and Vision: How Latecomers Came to Dominate Markets* (New York: McGraw Hill, 2001).

23 Stav Rosenzweig, "Non-Customers as Initiators of Radical Innovation," *Industrial Marketing Management* 66, no. 6 (2017): 1–12.

24 Our thesis suggests that the best defense against colonization by and territorial expanse of other nations is for a nation to stay at the cutting edge of innovation. To maintain sovereignty, a nation must either develop innovations first or quickly adopt and further develop innovations others have started.

25 Paul Kennedy, *The Rise and Fall of the Great Powers: Economic Change and Military Conflict from 1500 to 2000,*" (New York: Vintage Books, 1987); Daron Acemoglu and James A. Robins, *Why Nations Fail: The Origins of Power, Prosperity, and Poverty* (New York: Random House, 2012).

26 J. A. Schumpeter, *Business Cycles: A Theoretical, Historical and Statistical Analysis of the Capitalist Process* (New York: McGraw Hill, 1939); Christopher Freeman, "Innovation and Long Cycles of Economic Development," Paper presented at the International Seminar on Innovation and Development at the Industrial Sector, University of Campinas, 1982; Richard N. Foster, *Innovation: The Attacker's Advantage* (New York: Summit Books, 1986).

27 Immanuel Wallerstein, *The Modern World-System* (New York: Academic Press, 1974).

28 See, for example, Christopher Chase-Dunn and Peter Grimes, "World-System Analysis," *Annual Review of Sociology* 21 (1995): 387–417; Mauro F. Guillén and Sandra Suárez, "Explaining the Global Digital Divide: Economic, Political and Sociological Drivers of Cross-National Internet Use," *Social Forces* 84 (December 2005): 681–708; Thomas, D. Hall, *A World-Systems Reader: New*

Perspectives on Gender, Urbanism, Cultures, Indigenous Peoples, and Ecology (Lanham, MD: Rowman & Littlefield, 2000); Mathew C. Mahutga, "The Persistence of Structural Inequality? A Network Analysis of International Trade 1965–2000," *Social Forces* 84 (June 2006): 1863–89; John M. Shandra et al., "Multinational Corporations, Democracy and Child Mortality: A Quantitative, Cross-National Analysis of Developing Countries," *Social Indicators Research* 73 (2005): 267–93; Ronan Van Rossem, "The World System Paradigm as General Theory of Development: A Cross-National Test," *American Sociological Review* 61 (June 1996): 508–27.

29 Scholars seem to agree that there are individual cases of mobility in the hierarchy, but that overall, positions are stable with no large upward or downward shifts in power. See the review of Christopher Chase-Dunn and Peter Grimes, "World-System Analysis," *Annual Review of Sociology* 21 (1995): 387–417. See also Mauro F. Guillén, and Sandra Suárez, "Explaining the Global Digital Divide: Economic, Political and Sociological Drivers of Cross-National Internet Use," *Social Forces* 84 (December 2005): 681–708; Immanuel Wallerstein, *The Modern World-System* (New York: Academic Press, 1974).

2 Roman Concrete: Foundations of an Empire

1 Lynne C. Lancaster, *Concrete Vaulted Construction in Imperial Rome: Innovations In Context* (Cambridge: Cambridge University Press, 2005), 3.

2 David Moore, *The Roman Pantheon: The Triumph of Concrete* (Mangilao: University of Guam Station, 1995), 17.

3 Ibid.

4 V. M. Malhotra and Povindar K. Mehta, *Advances in Concrete Technology: Pozzolanic and Cementitious Materials*, vol. 1 (Amsterdam: Gordon and Breach Science Publishers, 1996), 3.

5 Moore, *Roman Pantheon*, 27.

6 George R. H. Wright, *Ancient Building Technology*, vol. 2, part 1 (Leiden: Brill Publishing, 2005), 192.

7 Ibid., 175.

8 Rabun M. Taylor, *Roman Builders: A Study in Architectural Process* (Cambridge: Cambridge University Press, 2003), 215–16.

9 Wright, *Ancient Building Technology*, 194.

10 Taylor, *Roman Builders*, 8.

11 Ann Kerns, *Seven Wonders of Architecture* (Minneapolis: Lerner Publishing Group, 2010).

12 Lancaster, *Concrete Vaulted Construction in Imperial Rome*, 5.

13 Moore, *Roman Pantheon*, 17.

14 Wright, *Ancient Building Technology*, 186.

15 Ibid., 210.

16 Ibid., 116.

17 Moore, *Roman Pantheon*, 18.

18 Janet Delaine, "Structural Experimentation: The Lintel Arch, Corbel and Tie in Western Roman Architecture," *World Archeology* 21 no. 3 (1990): 407–24.

19 Lancaster, *Concrete Vaulted Construction in Imperial Rome*, 65.

20 Taylor, *Roman Builders*, 8.

21 Lancaster, *Concrete Vaulted Construction in Imperial Rome*, 3.

22 Morris H. Morgan, *Vitruvius: The Ten Books on Architecture* (Cambridge: Harvard University Press, 1914), 46.

23 Pliny the Elder, *The Natural History*, trans. John Bostock (London: Taylor and Francis, 1855).

24 Lancaster, *Concrete Vaulted Construction in Imperial Rome*, 166.

25 Nancy Ramage and Andrew Ramage, *Roman Art*, 5th ed. (Upper Saddle River, NJ: Pearson Prentice Hall, 2009), 161.

26 Lancaster, *Concrete Vaulted Construction in Imperial Rome*, 17.

27 Ramage and Ramage, *Roman Art*, 32.

28 T. G. Tucker, *Life in the Roman World of Nero and St. Paul* (New York: Macmillan Company, 1910), 90.

29 Sir Banister Fletcher, *A History of Architecture*, 18th ed. (New York: Charles Scribner's Sons, 1975), 256–343; James E. Packer, *The Forum of Trajan in Rome: A Study of the Monuments in Brief* (London: University of California Press, 2001), 174–95; Ramage, *Roman Art*, 5th ed., 73–6, 113–20, 210, 223, 231, 235, 237, 239, 304, 318, 331, 333, 344 and 353–7; John W. Stamper, *The Architecture of Roman Temples: The Republic to the Middle Empire* (Cambridge: Cambridge University Press, 2005), 48; Marilyn Stokstad, *Art History: Ancient Art*, 3rd ed. (Upper Saddle River, NJ: Pearson Prentice Hall, 2008), 182; Jean Pierre Adam and Anthony Matthews, *Roman Building: Materials and Techniques* (Milton Park, Abingdon, Oxfordshire: Routledge, 2005), 58.

30 Ramage, *Roman Art*, 72–3.

31 Ibid., 74–5.

32 Stokstad, *Art History*, 183.

33 Ramage, *Roman Art*, 170.

34 Moore, *Roman Pantheon*, 17.

35 John Henry Middleton, *The Remains of Ancient Rome*, vol. 1 (London: Adam and Charles Black, 1892), 251; William Smith, *Dictionary of Greek and Roman Antiquities* (New York: Harper & Brothers, 1857), 380.

36 Stokstad, *Art History*, 182.

37 Moore, *Roman Pantheon*, 17.

38 Pliny, *Natural History*.

39 Wright, *Ancient Building Technology*, 204.

40 Arthur Segal, *Theatres in Roman Palestine and Provincia Arabia* (New York: E. J. Brill, 1995), 64.

41 William Smith, *Dictionary of Greek and Roman Antiquities* (London: Samuel Bentley, 1842), 617.

42 Ibid.

43 Lancaster, *Concrete Vaulted Construction in Imperial Rome*, 174.

44 Janet DeLaine, *The Baths of Caracalla: A Study in the Design, Construction, and Economics of Large-Scale Building Projects in Imperial Rome* (Portsmouth, RI: Journal of Roman Archaeology, 1997), 269.

45 Taylor, *Roman Builders*, 8–9.

46 Donald J. Conway, *Human Response to Tall Buildings* (Stroudsburg: Dowden, Hutchinson & Ross, 1977), 41.

47 Colin Szasz, "The Influence of Roman Engineering and Architecture." http://www.arch.mcgill.ca/prof/sijpkes/arch304/winter2001/cszasz/u1/roman.htm.

48 Tractate Shabbat, *Babylonian Talmud,* 33.

49 Morgan, *Vitruvius,* 56.

50 Ramage, *Roman Art,* 32.

51 Stokstad, *Art History,* 224.

52 Tucker, *Life in the Roman World of Nero and St. Paul,* 79.

53 Lancaster, *Concrete Vaulted Construction in Imperial Rome,* 174.

54 Tucker, *Life in the Roman World of Nero and St. Paul,* 424.

55 Bryan Ward-Perkins, *From Classical Antiquity to the Middle Ages: Urban Public Building in Northern and Central Italy, CE 300–850* (New York: Oxford University Press, 1984), 4–6.

56 Geoffrey S. Sumi, *Ceremony and Power: Performing Politics in Rome between Republic and Empire* (Ann Arbor: University of Michigan Press, 2005), 268.

57 Segal, *Theatres in Roman Palestine and Provincia Arabia,* 4.

58 Stokstad, *Art History,* 170.

59 Lesley Adkins and Roy A. Adkins, *Handbook to Life in Ancient Rome* (New York: Facts On File, 2004), 213.

60 Edwin W. Bowen, "Roman Commerce in the Early Empire," *Classical Weekly* 21 no. 26 (May 14, 1928): 201–6.

61 Adkins, *Handbook to Life in Ancient Rome,* 215.

62 Ying-shih Yü, *Trade and Expansion in Han China: A Study in Structure of Sino-Barbarian Economic Relations* (Berkeley: University of California Press, 1967), 159.

63 Stokstad, *Art History,* 224.

64 George Mousourakis, *A Legal History of Rome* (New York: Routledge, 2007), 22.

65 Ibid.

66 Ibid., 10.

67 Charlotte Elizabeth Goodfellow, *Roman Citizenship: A Study of Its Territorial and Numerical Expansion from the Earliest Times to the Death of Augustus* (Lancaster: Lancaster Press, 1935), 14.

68 Mousourakis, *A Legal History of Rome,* 199.

69 Goodfellow, *Roman Citizenship: A Study of Its Territorial and Numerical Expansion,* 20.

70 A. N. Sherwin-White, *The Roman Citizenship* (Oxford: Oxford University Press, 1973), 148.

71 Goodfellow, *Roman Citizenship: A Study of Its Territorial and Numerical Expansion,* 22.

72 Sherwin-White, *The Roman Citizenship,* 165.

73 Ibid., 225.

74 Ibid., 280.

75 Mousourakis, *A Legal History of Rome,* 11.

76 Anthony R. Birley, *Septimius Severus: The African Emperor* (New York: Routledge, 1988), xi.

77 Patricia Southern, *The Roman Empire from Severus to Constantine* (New York: Routledge, 2001), 64.

78 Ibid.

79 DeLaine, "The Baths of Caracalla," 269.

80 Moore, *The Roman Pantheon*, 166.

81 DeLaine, "The Baths of Caracalla," 269.

82 Moore, *The Roman Pantheon*, 168.

83 Stokstad, *Art History*, 170.

84 Wright, *Ancient Building Technology*, 215.

85 Lancaster, *Concrete Vaulted Construction in Imperial Rome*, 18–21.

86 Ibid., 178.

87 Adkins, *Handbook to Life in Ancient Rome*, 115–6.

88 Some examples include Arthur E. R. Boak, *Manpower Shortage and the Fall of the Roman Empire in the West* (Westport, CT: Greenwood Press, 1955); Arther Ferril, *The Fall of the Roman Empire: The Military Explanation* (London: Thames and Hudson, 1986); Edward Gibbon, *The Decline and Fall of the Roman Empire* (Suffolk: Richard Clay and Company, 1960); Michael I. Rostovtzeff, "The Empire During the Anarchy," in *The End of the Roman Empire: Decline or Transformation?*, ed. Donald Kagan (Lexington: D. C. Heath and Company, 1992), 29–39; William H. McNeill, *A World History* (New York: Oxford University Press, 1999).

3 Swift Equine Warfare and the Rise of Mongol Power

1 Morris Rossabi, *The Mongols: A Very Short Introduction* (New York: Oxford University Press, 2012).

2 Jack Weatherford, *Genghis Khan and the Making of the Modern World* (New York: Broadway Books, 2005); Timothy May, *The Mongol Conquests in World History* (London: Reaktion Books, 2012).

3 Weatherford, *Genghis Khan and the Making of the Modern World*; May, *Mongol Conquests in World History*.

4 Weatherford, *Genghis Khan and the Making of the Modern World*; May, *Mongol Conquests in World History*.

5 Stephen R. Turnbull, *Essential Histories: Genghis Khan & the Mongol Conquests 1190–1400* (New York: Routledge, 2004).

6 Turnbull, *Essential Histories*.

7 Morris Rossabi, *All the Khan's Horses* (New York: Natural History Magazine, 1994).

8 Weatherford, *Genghis Khan and the Making of the Modern World*.

9 Ibid.

10 Ibid.

11 Robert Drews, *Early Riders: The Beginnings of Mounted Warfare in Asia and Europe* (New York: Routledge, 2004).

12 Phillip Sidnell, *Warhorse: Cavalry in Ancient Warfare* (New York: Hambledon Continuum, 2006).

13 Drews, *Early Riders*.

14 Ibid.

15 M. Polo, *The Travels of Marco Polo*, 2 vols., trans. Henry Yule, ed. Henri Cordier (New York: Dover Publications, 1993), 262.

16 Turnbull, *Essential Histories*.

17 Drews, *Early Riders*.

18 May, *Mongol Conquests in World History*.

19 Urgunge Onon, *The Secret History of the Mongols: The Life and Times of Chinggis Khan*, new ed. (London and New York: Routledge Curzon, 2001).

20 Rossabi, *The Mongols*.

21 Ibid.

22 Sidnell, *Warhorse*.

23 Weatherford, *Genghis Khan and the Making of the Modern World*.

24 George Lane, *Genghis Khan and Mongol Rule* (Westport, CT: Greenwood, 2004), 31.

25 May, *Mongol Conquests in World History*.

26 Ibid.

27 Rossabi, *The Mongols*.

28 Drews, *Early Riders*.

29 John Morton, Simon Morton and Nicholas Morton, *Crusading and Warfare in the Middle Ages: Realities and Representations* (Burlington, VT: Ashgate Publishing, 2014).

30 Sidnell, *Warhorse*.

31 May, *Mongol Conquests in World History*.

32 Drews, *Early Riders*.

33 Richard Gabriel, "The Right Hand of Khan," *Military History* 25 no. 2 (2008): 42–9.

34 Turnbull, *Essential Histories*.

35 Drews, *Early Riders*.

36 Turnbull, *Essential Histories*.

37 Reuven Amitai-Preis, *Mongols and Mamluks: The Mamluk-Īlkhānid War, 1260–1281* (Cambridge: Cambridge University Press, 1995), 222.

38 John Keegan, *A History of Warfare* (New York: Random House, 2011).

39 Turnbull, *Essential Histories*.

40 May, *Mongol Conquests in World History*.

41 Onon, *Secret History of the Mongols*.

42 Turnbull, *Essential Histories*.

43 May, *Mongol Conquests in World History*.

44 Ibid.

45 Turnbull, *Essential Histories*.

46 May, *Mongol Conquests in World History*.

47 Ibid.

48 Turnbull, *Essential Histories*.

49 Weatherford, *Genghis Khan and the Making of the Modern World*.

50 Turnbull, *Essential Histories*.
51 Peter Jackson, *The Mongols and the West, 1221–1410* (New York: Routledge, 2014).
52 Weatherford, *Genghis Khan and the Making of the Modern World*.
53 May, *Mongol Conquests in World History*.
54 Turnbull, *Essential Histories*.
55 Roland Oliver and Anthony Atmore, *Medieval Africa, 1250–1800* (Cambridge: Cambridge University Press, 2001), 17.
56 May, *Mongol Conquests in World History*.
57 Drews, *Early Riders*.
58 Ibid.
59 Onon, *Secret History of the Mongols*.
60 Drews, *Early Riders*.
61 Ibid.
62 James E. Lindsay, *Daily Life in the Medieval Islamic World* (Westport, CT: Greenwood Press, 2005).
63 Turnbull, *Essential Histories*.
64 May, *Mongol Conquests in World History*.
65 Ibid.
66 Ibid.
67 Turnbull, *Essential Histories*.
68 Ibid.
69 Ibid.
70 Ibid.
71 May, *Mongol Conquests in World History*.
72 Weatherford, *Genghis Khan and the Making of the Modern World*.
73 May, *Mongol Conquests in World History*.
74 John Man, *The Mongol Empire: Genghis Khan, His Heirs and the Founding of Modern China* (London: Transworld Publishers, 2014).
75 Weatherford, *Genghis Khan and the Making of the Modern World*.
76 May, *Mongol Conquests in World History*.
77 Ibid.
78 Ibid.
79 Ibid.
80 Ibid.
81 Ibid.
82 Ibid.
83 Turnbull, *Essential Histories*.
84 May, *Mongol Conquests in World History*.
85 Ibid.
86 May, *Mongol Conquests in World History*.
87 Ibid.
88 Ibid.
89 McLynn, *Genghis Khan*.
90 Ibid.
91 Man, *Mongol Empire*.
92 May, *Mongol Conquests in World History*.

93 Man, *Mongol Empire*.

94 May, *Mongol Conquests in World History*.

95 Turnbull, *Essential Histories*.

96 Ibid.

97 Man, *Mongol Empire*.

98 May, *Mongol Conquests in World History*.

99 Frank McLynn, *Genghis Khan: His Conquests, His Empire, His Legacy* (Boston: Da Capo Press, 2015).

100 May, *Mongol Conquests in World History*.

101 Man, *Mongol Empire*.

102 McLynn, *Genghis Khan*.

103 May, *Mongol Conquests in World History*.

104 Ibid.

105 Man, *Mongol Empire*.

106 May, *Mongol Conquests in World History*.

4 How Gunpowder Shaped the Fortunes of Nations

1 J. P. A. Agrawal, *High Energy Materials: Propellants, Explosives and Pyrotechnics* (Hoboken, NJ: Wiley-VCH (Imprint) John Wiley & Sons, 2010).

2 T. Andrade, *The Gunpowder Age: China, Military Innovation, and the Rise of the West in World History* (Princeton, NJ: Princeton University Press, 2016), 31.

3 Joseph Needham et al., *Science and Civilization in China*, vol. 5, no. 7, *Military Technology: The Gunpowder Epic* (Cambridge: Cambridge University Press), 149.

4 Ibid., 161–210.

5 Ibid., 276 and 472.

6 Ibid., 108.

7 L. Wu, T. L. Davis and W. Po-Yang, "An Ancient Chinese Treatise on Alchemy Entitled Ts'an T'ung Ch'I," *Isis* 18, no. 2 (1932): 210–89.

8 Brenda J. Buchanan, ed., *Gunpowder, Explosives and the State: A Technological History* (Aldershot: Ashgate, 2006), 2.

9 Andrade, *Gunpowder Age*, 31.

10 Needham et al., *Science and Civilization in China*, 147.

11 Andrade, *Gunpowder Age*, 32.

12 Ibid., 24–25.

13 Ibid., 32.

14 Needham et al., *Science and Civilization in China*, 211.

15 Ibid.

16 Andrade, *Gunpowder Age*, 32.

17 Needham et al., *Science and Civilization in China*, 171.

18 Andrade, *Gunpowder Age*, 41.

19 Needham et al., *Science and Civilization in China*, 163.

20 Ibid., 128–29.

21 Andrade, *Gunpowder Age*, 35.

22 Stephen G. Haw, "The Mongol Empire: The First 'Gunpowder Empire',"
   Journal of the Royal Asiatic Society 23, no. 3 (2013): 441–69.
23 Thomas T. Allsen, "The Circulation of Military Technology in the Mongolian
   Empire," in *Warfare in Inner Asian History (500–1800)*, ed. Nicola De Cosmo
   (Leiden: Brill, 2002), 276–77.
24 Andrade, *Gunpowder Age*, 45–47.
25 Ibid.
26 L. Carrington Goodrich, *A Short History of the Chinese People* (Mineola,
   NY: Dover, 2002), 173.
27 Needham et al., *Science and Civilization in China*, 574.
28 Ibid., 232.
29 Ibid., 236; Janet L. Abu-Lughod, *Before European Hegemony: The World System
   A.D. 1250–1350* (New York: Oxford University Press, 1989), 326.
30 Needham et al., *Science and Civilization in China*, 294.
31 Andrade, *Gunpowder Age*, 53.
32 Ibid.
33 Needham et al., *Science and Civilization in China*, 263–76.
34 Ibid., 9.
35 Andrade, *Gunpowder Age*, 55.
36 Needham et al., *Science and Civilization in China*, 314–41.
37 Ibid., 568–79.
38 Kenneth Chase, *Firearms: A Global History to 1700* (Cambridge;
   New York: Cambridge University Press, 2003), 58.
39 Needham et al., *Science and Civilization in China*, 573.
40 Chase, *Firearms: A Global History to 1700*, 58.
41 Needham et al., *Science and Civilization in China*, 577.
42 Walter de Milemete called De Nobilitatibus, sapientii et prudentiis regum.
43 Andrade, *Gunpowder Age*, 76.
44 Claude Blair, "The Milemete Guns," *Journal of the Ordnance Society* 16
   (2004): 5–16.
45 John M. Hobson, *The Eastern Origins of Western Civilization* (Cambridge:
   Cambridge University Press, 2004), 187.
46 Chase, *Firearms: A Global History to 1700*, 59.
47 Ibid., 60.
48 J.R. Partington, *A History of Greek Fire and Gunpowder* (Baltimore: Johns
   Hopkins University Press, 1960), 104.
49 Andrade, *Gunpowder Age*, 78.
50 Philippe Contamine, *War in the Middle Ages* (New York: Basil Blackwell,
   1984), 139.
51 Andrade, *Gunpowder Age*, 83.
52 Ibid., 88.
53 Jean Froissart, *Chronigues de J. Froissart, tome huitieme, 1370–1377, premiere partie*,
   ed. Simeon Luce (Paris: Librairie Renouard, 1888), 279, cited in Andrade,
   Gunpowder Age, 90.
54 Andrade, *Gunpowder Age*, 3.

55 Clifford J. Rogers, "The Military Revolutions of the Hundred Years' War," *Journal of Military History*, 57, no. 2 (1993): 241–78.

56 Kelly DeVries, "The Use of Gunpowder Weaponry by and against Joan of Arc during the Hundred Years War," *War and Society* 15, no. 1 (1996): 1–15.

57 Andrade, *Gunpowder Age*, 92.

58 Ibid.

59 Henry Brackenbury, *Ancient Cannon in Europe, Part II: From CE 1351 to CE 1400* (Woolwich: Royal Artillery Institution, 1866), 39.

60 Andrade, *Gunpowder Age*, 91.

61 Chase, *Firearms: A Global History to 1700*, 95.

62 Iqtidar Alam Khan, "Coming of Gunpowder to the Islamic World and North India: Spotlight on the Role of the Mongols," *Journal of Asian History* 30, no. 1 (1996): 27–45.

63 Ahmad Y. al-Hassan and Donald R. Hill, *Islamic Technology: An Illustrated History* (Cambridge: Cambridge University Press, 1992), 106–20.

64 Needham et al., *Science and Civilization in China*, 39–42.

65 Gabor Agoston, "Firearms and Military Adaptation: The Ottomans and the European Military Revolution, 1450–1800," *Journal of World History* 25, no. 1 (2014): 88.

66 Ibid., 89.

67 Douglas E. Streusand, *Islamic Gunpowder Empires: Ottomans, Safavids, and Mughals* (Boulder, CO: Westview Press, 2010), 41.

68 Franz Babinger, *Mehmed the Conqueror and His Time* (Princeton, NJ: Princeton University Press, 1978), 449.

69 Andrade, *Gunpowder Age*, 94.

70 Agoston, "Firearms and Military Adaptation," 109.

71 Andrade, *Gunpowder Age*.

72 Kelly DeVries and Robert Douglas Smith, "Breech-loading Guns with Removable Powder Chambers: A Long-lived Military Technology," in *Gunpowder, Explosives and the State: A Technological History*, ed. Brenda J. Buchanan (Aldershot; Burlington: Ashgate, 2006), 252.

73 Bert S. Hall, "The Corning of Gunpowder and the Development of Firearms in the Renaissance," in *Gunpowder: The History of an International Technology*, ed. Brenda J. Buchanan (Bath University Press, 1996), 91.

74 Chase, *Firearms: A Global History to 1700*, 60.

75 Hall, "The Corning of Gunpowder and the Development of Firearms in the Renaissance," 100–3.

76 Roger M. Savory, "The Sherley Myth," *Iran: Journal of the British Institute of Persian Studies* 5 (1967): 73–81, in *Studies on the History of Safavid Iran* (London: Variorum, 1987).

77 J.L. Bacqué-Grammont, *Les Ottomans, les Safavides et leurs voisins: Contribution à l'histoire des relations internationales dans l'Orient islamique de 1514 à 1524* (Istanbul and Leiden: 1987), 165, as cited in *Encyclopaedia Iranica*, online edition, New York, 1996.

78 Chase, *Firearms: A Global History to 1700*, 117.

79 B. Scarcia Amoretti, ed., *Šāh Ismāʿīl I nei "Diarii" di Marin Sanudo I* (Rome, 1979), as cited in *Encyclopaedia Iranica*, online edition, New York, 1996.

80 Streusand, *Islamic Gunpowder Empires*, 169.
81 Chase, *Firearms: A Global History to 1700*, 119.
82 Ibid., 121.
83 Ibid., 122.
84 Streusand, *Islamic Gunpowder Empires*, 212.
85 Ibid., 255.
86 Chase, *Firearms: A Global History to 1700*, 132.
87 Ibid.
88 Ibid.
89 Annemarie Schimmel, *The Empire of the Great Mughals: History, Art and Culture* (London: Reaktion Books, 2004), 88.
90 Streusand, *Islamic Gunpowder Empires*, 257.
91 Jos Gommans, *Mughal Warfare: Indian Frontiers and High Roads to Empire, 1500–1700* (London: Routledge: 2002), 134.
92 Streusand, *Islamic Gunpowder Empires*.
93 P. Jakov Smith, "Eurasian Transformations of the Tenth to Thirteenth Centuries: The View from Song China, 960–1279," *Medieval Encounters* 10, no. 1–3 (2004): 279–308.
94 Andrade, *Gunpowder Age*.
95 Ibid.
96 Ibid.
97 Frank Tallett, *War and Society in Early Modern Europe* (London: Routledge, 1997).
98 Andrade, *Gunpowder Age*, 93.
99 Mario Philippides and Walter K. Hanak, *The Siege and the Fall of Constantinople in 1453: Historiography, Topography, and Military Studies* (Farnham Surrey: Ashgate, 2011).
100 Andrade, *Gunpowder Age*, 94.
101 Leonardo di Chio, "Letter to Pope Nicholas V," August 16, 1453, ed. J. P. Migne, *Patrologia Graeca* 159, 923A–944B.
102 Gabor Agoston, "Ottoman Artillery and European Military Technology in the Fifteenth and Seventeenth Centuries," *Acta Orientalia Academiae Scientiarum Hungaricae* 47, no. 1/2 (1994): 28.
103 Noel Perrin, *Giving Up the Gun: Japan's Reversion to the Sword, 1543–1879* (Boston: David R. Godine, Publisher, 1979).
104 Daron Acemoglu and James A. Robinson, "Economic Backwardness in Political Perspective," *American Political Science Review* 100, no. 1 (2006): 115–31.

5 Golden Age of Chinese Water Navigation

1 Edward L. Dwyer, *China and the Oceans in the Early Ming Dynasty* (London: Pearson Education, 2006); Louise Levathes, *When China Ruled the Seas: The Treasure Navy of the Dragon Throne, 1405–1433* (New York: Oxford University Press, 1996); Gavin Menzies, *1421: The Year China Discovered America* (2002).

2 Dwyer, *China and the Oceans in the Early Ming Dynasty*; Kuei-Sheng Chang, "The Maritime Scene in China at the Dawn of Great European Discoveries," *Journal of the American Oriental Society* 94, no. 3 (1974): 348–59; Levathes, *When China Ruled the Seas*, 78–85.

3 Dwyer, *China and the Oceans in the Early Ming Dynasty*; Levathes, *When China Ruled the Seas*, 78–85; Chang, "The Maritime Scene in China at the Dawn of Great European Discoveries," 348–59; Evan Hadingham, "Ancient Chinese Explorers," PBS NOVA, January 16, 2001, http://www.pbs.org/wgbh/nova/ancient/ancient-chinese-explorers.html.

4 Dwyer, *China and the Oceans in the Early Ming Dynasty*.

5 Ibid.

6 Levathes, *When China Ruled the Seas*, 78–85; Robert Temple, *The Genius of China: 3,000 Years of Science, Discovery & Invention* (Rochester: Inner Traditions, 2007), 250–54; Dwyer, *China and the Oceans in the Early Ming Dynasty*.

7 Dwyer, *China and the Oceans in the Early Ming Dynasty*; Levathes, *When China Ruled the Seas*, 78–85.

8 Menzies, *1421: The Year China Discovered America*.

9 Levathes, *When China Ruled the Seas*.

10 For more on the Yongle emperor (1360–1424), the third emperor of the Ming dynasty, see Shih-Shah Henry Tsai, *Perpetual Happiness: The Ming Emperor Yongle* (Seattle: University of Washington Press, 2001).

11 Menzies, *1421: The Year China Discovered America*.

12 Ibid.

13 Levathes, *When China Ruled the Seas*.

14 Ibid.

15 Daron Acemoglu and James A. Robinson, *Why Nations Fail: The Origins of Power, Prosperity and Poverty* (New York: Crown Business, 2012).

16 For more on Confucianism, see Yao Xinzhong, *An Introduction to Confucianism* (Cambridge: Cambridge University Press, 2000).

17 Andreas Bøje Forsby, "The Non-Western Challenger? The Rise of a Sino-Centric China," *DIIS Report* 16 (2011): 24–27.

18 Levathes, *When China Ruled the Seas*.

19 Gang Deng, *Chinese Maritime Activities and Socioeconomic Development, c. 2100 BC–1900 CE* (Westport, CT: Greenwood Press, 1997)

20 Levathes, *When China Ruled the Seas*.

21 Menzies, *1421: The Year China Discovered America*.

22 Ibid.

23 Dwyer, *China and the Oceans in the Early Ming Dynasty*, 109.

24 Levathes, *When China Ruled the Seas*.

25 Biography of Zheng He, as reported in Dwyer, *China and the Oceans in the Early Ming Dynasty*.

26 Levathes, *When China Ruled the Seas*.

27 Dwyer, *China and the Oceans in the Early Ming Dynasty*.

28 Ibid.

29 Gerard J. Tellis, *Unrelenting Innovation* (San Francisco: Jossey-Bass, 2013).

30 Deng, *Chinese Maritime Activities and Socioeconomic Development*, 8–14; Dwyer, *China and the Oceans in the Early Ming Dynasty*.

31 Johns King Fairbank and Merle Goldman, *China: A New History* (Cambridge, MA: Belknap Press of Harvard University Press, 1998), 138–40.

32 Eric Mielants, *The Origins of Capitalism and the "Rise of the West"* (Philadelphia: Temple University Press, 2007), 67.

33 Dwyer, *China and the Oceans in the Early Ming Dynasty*.

34 Levathes, *When China Ruled the Seas*.

35 Dwyer, *China and the Oceans in the Early Ming Dynasty*.

36 Ibid.

37 Ibid.

38 Ibid.

39 Ibid.

40 Menzies, *1421: The Year China Discovered America*.

6 Venetian Shipbuilding: Mastering the Mediterranean

1 Roger Crowley, *City of Fortune: How Venice Ruled the Seas* (New York: Random House Trade Paperbacks, 2011).

2 Ibid.

3 Benjamin Arbel, "Venice's Maritime Empire in the Early Modern Period," in *A Companion to Venetian History, 1400–1797*, ed. Eric R. Dursteler (Leiden: Brill, 2013), 125–253.

4 Nükhet Varlik, *Plague and Empire in the Early Modern Mediterranean World: The Ottoman Experience, 1347–1600* (Cambridge: Cambridge University Press, 2015).

5 Frederick C. Lane, *Venetian Ships and Shipbuilders of the Renaissance* (New York: Orno Press, 1979), 7.

6 Robert C. Davis, "Shipbuilders of the Venetian Arsenal: Workers and Workplace in the Preindustrial City" (Baltimore: Johns Hopkins University Press, 1991), 15.

7 Frederick C. Lane, *Venice: A Maritime Republic* (Baltimore: John Hopkins University Press, 1973), 13.

8 William H. McNeill, *Venice: The Hinge of Europe, 1081–1797* (Chicago: University of Chicago Press, 1974), 10; Lane, *Venice*, 8.

9 McNeill, *Venice: The Hinge of Europe, 1081–1797*.

10 Ibid., 11.

11 Elizabeth Horodowich, *A Brief History of Venice* (London: Constable & Robinson, 2009).

12 Davis, "Shipbuilders of the Venetian Arsenal," 11.

13 Ibid., 12.

14 McNeill, *Venice: The Hinge of Europe, 1081–1797*, 6.

15 Davis, "Shipbuilders of the Venetian Arsenal," 15.

16 John H. Pryor, *Geography, Technology and War: Studies in the Maritime History of the Mediterranean, 649–1571* (Cambridge: Cambridge University Press, 1988).

17 Diego Puga and Daniel Trefler, "International Trade and Institutional Change: Medieval Venice's Response to Globalization," Working Paper 18288, NBER, 2012.

18 Puga and Trefler, "International Trade and Institutional Change."

19 Daron Acemoglu and James Robinson, *Why Nations Fail: The Origins of Power, Prosperity, and Poverty* (New York: Crown Publishers, 2012).

20 Ibid.

21 Crowley, *City of Fortune*; Puga and Treflor, "International Trade and Institutional Change."

22 Lane, *Venice: A Maritime Republic*; Puga and Trefler, "International Trade and Institutional Change."

23 Puga and Trefler, "International Trade and Institutional Change"; Acemoglu and Robinson, *Why Nations Fail*.

24 Crowley, *City of Fortune*; Timothy E. Gregory, *A History of Byzantium* (Malden, MA: John Wiley & Sons, 2010); Robert Bideleux and Ian Jeffries, *A History of Eastern Europe: Crisis and Change* (London: Routledge, 1998); Olivia Remie Constable, *Housing the Stranger in the Mediterranean World: Lodging, Trade, and Travel in Late Antiquity and the Middle Ages* (Cambridge: Cambridge University Press, 2003).

25 Puga and Trefler, "International Trade and Institutional Change".

26 Ibid., 2.

27 Crowley, *City of Fortune*, 17.

28 Ibid.

29 Ibid., 19.

30 Thomas F. Madden, *Venice: A New History* (New York: Penguin Books, 2013), 117.

31 Crowley, *City of Fortune*, 118.

32 Lane, *Venice: A Maritime Republic*, 196.

33 Ibid., 128.

34 Ibid., 198.

35 Ibid.

36 Ibid., 236.

37 Ibid.

38 Lane, *Venetian Ships and Shipbuilders of the Renaissance*.

39 Riccardo Calmani, *The Venetian Ghetto: The History of a Persecuted Community* (Amazon Digital Services, 2013).

40 Calmani, *The Venetian Ghetto*.

41 Lane, *Venice: A Maritime Republic*.

42 Ibid., 152.

43 Ibid.

44 This story is from Puga and Trefler, "International Trade and Institutional Change."

45 Katerina Konstantinidou and Elpis Mantadakis, "Venetian Rule and Control of Plague Epidemics on the Ionian Islands during 17th and 18th Centuries," *Emerging Infectious Diseases* 15, no. 1 (2009): 39–43.

46 Davis, "Shipbuilders of the Venetian Arsenal," 94.

47 Ibid.
48 Lane, *Venice: A Maritime Republic*, 148.
49 Ibid., 164.
50 John Martin, *Organizational Behavior and Management* (London: Thomson Learning, 2005), 45.
51 Davis, "Shipbuilders of the Venetian Arsenal," 63.
52 Ibid., 68.
53 Ibid.
54 Ibid., 69.
55 Ibid., 68.
56 Ibid., 76.
57 Crowley, *City of Fortune*.
58 Ibid.
59 Armando Cortesao, *The Suma Oriental of Tomé Pires, 1512–1515* (London: Routledge, 2010), as reported in Crowley, *City of Fortune*, 374.
60 Acemoglu and Robinson, *Why Nations Fail*; Puga and Trefler, "International Trade and Institutional Change"; Benjamin Ravid, "Levantine and Ponentine Jewish Merchants: Venice, and Trade with the Ottoman Empire in the Late Sixteenth and Early Seventeenth Centuries," in *Studies in Jewish History Presented to Joseph Hacker*, ed. Yaron Ben-Naeh, Jeremy Cohen, Moshe Idel and Yosef Kaplan (Jerusalem: Zalman Shazar Center, 2014), 381–405.
61 Ravid, "Levantine and Ponentine Jewish Merchants."
62 Ibid.

7 Portuguese Caravel: Building an Oceanic Empire

1 William Hardy McNeill, *The Rise of the West: A History of the Human Community* (Chicago: University of Chicago, 1963), 570.
2 Margaret L. King, *Renaissance Humanism: An Anthology of Sources* (Indianopolis: Hackett Publishing Company, 2014), 321.
3 Kenneth Maxwell, "Portugal, Europe, and the Origins of the Atlantic Commercial System 1415–1520," *Portuguese Studies* 8 (1992): 5.
4 Ibid., 6.
5 Roger Crowley, *City of Fortune: How Venice Ruled the Seas* (New York: Random House, 2013).
6 Paolo Bernardini and Norman Fiering, *The Jews and Expansion of Europe to the West, 1450 to 1800* (New York: Berghahn Books, 2001), 478.
7 Fred Bronner, "Portugal and Columbus: Old Drives in New Discoveries," *Mediterranean Studies* 6 (1996): 56–57.
8 Gavin Menzies, *1421: The Year China Discovered America* (London: Transworld Publishers, 2002).
9 John M. Hobson, *The Eastern Origins of Western Civilization* (Cambridge: Cambridge University Press, 2004), 141.
10 Francisco Bethencourt, *Portuguese Oceanic Expansion: 1400–1800* (Cambridge: Cambridge University Press, 2007).

11 Menzies, *1421: The Year China Discovered America,* 347.
12 Bailey W. Diffie and George D. Winius, *Foundations of the Portuguese Empire: 1415–1580* (Minneapolis: University of Minnesota Press, 1977), 215–16.
13 Ibid.
14 Peter Barber et al., *Mapping Our World: Terra Incognita to Australia* (Canberra, Australia: National Library of Australia, 2013).
15 Hobson, *Eastern Origins of Western Civilization,* 142.
16 Bethencourt, *Portuguese Oceanic Expansion,* 472.
17 Diffie and Winius, *Foundations of the Portuguese Empire,* 140–41.
18 Bethencourt, *Portuguese Oceanic Expansion,* 484.
19 McNeill, *Rise of the West,* 538.
20 Bethencourt, *Portuguese Oceanic Expansion,* 485.
21 McNeill, *Rise of the West,* 539 and 547.
22 Joseph Needham, *Science and Civilization in China,* vol. 3, *Mathematics and the Sciences of the Heavens and the Earth* (Cambridge: Cambridge University Press, 1959), 20.
23 Bailey W. Diffie and George D. Winius, *Foundations of the Portuguese Empire: 1415-1580* (Minneapolis, Minnesota: University of Minnesota, 1977), 14–17 and 41.
24 Malyn Newitt, *A History of Portuguese Overseas Expansion 1400–1668* (New York: Routledge, 2005).
25 Diffie and Winius, *Foundations of the Portuguese Empire,* 25–26.
26 Hobson, *Eastern Origins of Western Civilization,* 180.
27 Menzies, *1421: The Year China Discovered America,* 341.
28 Hobson, *Eastern Origins of Western Civilization,* 141.
29 Roger Crowley, *Conquerors: How Portugal Forged the First Global Empire* (New York: Random House, 2015).
30 Crowley, *Conquerors.*
31 Crowley, *Conquerors*; Menzies, *1421.*
32 Bronner, "Portugal and Columbus," 57.
33 Angus Madison, *The World Economy: A Millennial Perspective* (Paris: Development Center of the Organization for Economic Co-Operation and Development (OECD), 2001).
34 Vitorino M. Haes, "Portugal and the Making of the Atlantic World: Sugar Fleets and Gold Fleets in the Seventeenth to the Eighteenth Centuries," *Fernand Braudel Center Review* 28 (2005): 314.
35 Ibid.
36 McNeill, *Rise of the West.*
37 Maxwell, "Portugal, Europe, and the Origins of the Atlantic Commercial System," 8.
38 Ibid.
39 Ibid., 5.
40 Ibid.
41 Haes, "Portugal and the Making of the Atlantic World," 314.

42 Maxwell, "Portugal, Europe, and the Origins of the Atlantic Commercial System," 5.
43 Hobson, *Eastern Origins of Western Civilization*, 136.
44 Jane S. Gerber, *The Jews of Spain: A History of the Sephardic Experience* (New York: Free Press, 1992).
45 James D. Tracy, *The Rise of Merchant Empires: Long Distance Trade in the Early Modern World: 1350–1750* (Cambridge: Cambridge University Press, 1990), 175
46 Thomas J. Misa, *Leonardo to the Internet: Technology and Culture from the Renaissance to the Present* (Baltimore: Johns Hopkins University Press, 2004), 39.
47 Ibid., 40.

8 The *Fluyt* and the Building of the Dutch Empire

1 Samuel J. Barkin, *Social Construction and the Logic of Money: Financial Predominance and International Economic Leadership* (Albany: State University of New York Press, 2003), 47.
2 Andrew Phillips and J. C. Sharman, *International Order in Diversity: War, Trade, and Rule in the Indian Ocean* (Cambridge: Cambridge University Press, 2015), 102.
3 Jan De Vries and A. M. van der Woude, *The First Modern Economy: Success, Failure, and Perseverance of the Dutch Economy, 1500–1815* (Cambridge: Cambridge University Press, 1997), 296.
4 Karel Davids, *The Rise and Decline of Dutch Technological Leadership* (Leiden: Brill Publishing, 2008), 95.
5 Wendy Van Duivenvoorde, *Dutch East India Company Shipbuilding: The Archaeological Study of Batavia and Other Seventeenth Century VOC Ships* (College Station: Texas A&M University Press, 2015).
6 Claudia Rei, *Turning Points in Leadership: Shipping Technology in the Portuguese and Dutch Merchant Empires* (Nashville: Vanderbilt University Press, 2011).
7 Duivenvoorde, *Dutch East India Company Shipbuilding*.
8 Jan De Vries, *The Economy of Europe in an Age of Crisis, 1600–1750* (Cambridge: Cambridge University Press, 1976), 118.
9 Davids, *Rise and Decline of Dutch Technological Leadership*, 99.
10 Ibid., 100.
11 Robert Parthesius, *Dutch Ships in Tropical Waters: The Development of the Dutch East India Company* (Amsterdam: Amsterdam University Press, 2010), 34.
12 Charles H. Parker, *Global Interactions in the Early Modern Age, 1400–1800* (Cambridge: Cambridge University Press, 1958), 24.
13 De Vries, *Economy of Europe in an Age of Crisis*, 118; Jonathan Israel, *The Dutch Republic: Its Rise Greatness and Fall 1477–1806* (Oxford: Clarendon Press).
14 De Vries and Woude, *First Modern Economy*, 72.
15 Willem Maas, "Immigrant Integration, Gender, and Citizenship in the Dutch Republic," *Politics, Groups, and Identities* 1, no. 3 (July 2013): 390–401.

16 J. Israel, "A Golden Age: Innovation in Dutch Cities, 1648–1720," *History Today*, March 3, 1995, 14.

17 Oscar Gelderblom, "Chapter 6: The Golden Age of the Dutch Republic," in *The Invention of Enterprise: Entrepreneurship from Ancient Mesopotamia to Modern Times*, ed. David S. Landes, Joel Mokyr, and William J. Baumol (Princeton, NJ: Princeton University Press, 2010), 156–82.

18 J. van Lottum, *Close Encounters with the Dutch: Across the North Sea: The Impact of the Dutch Republic on International Labour Migration, c. 1550–1850* (Amsterdam: Amsterdam University Press, 2008).

19 Gelderblom, "Golden Age of the Dutch Republic."

20 Anne E. C. McCants, *Civic Charity in a Golden Age: Orphan Care in Early Modern Amsterdam* (Urbana: University of Illinois Press, 1997).

21 Gelderblom, "Golden Age of the Dutch Republic."

22 Ibid.

23 Ibid, 169–70.

24 J. L. Price, *Dutch Culture in the Golden Age* (London: Reaktion Books, 2012).

25 Danielle van den Heuvel, *Women and Entrepreneurship: Female Traders in the Northern Netherlands, c. 1580–1815* (Amsterdam: Aksant, 2007).

26 Ariadne Schmidt, "Women and Guilds: Corporations and Female Labour Market Participation in Early Modern Hollan," *Gender and History* 21 no. 1 (2009): 170–89.

27 Ibid.

28 Jan De Vries, *Dutch Rural Economy* (New Haven, CT: Yale University Press, 1974), 49–56.

29 De Vries and Woude, *First Modern Economy*.

30 *The Political Economy of the Dutch Republic*, ed. Oscar Gelderblom (Abingdon: Routledge, 2016), 174.

31 Ibid., 177.

32 Gelderblom, "Golden Age of the Dutch Republic."

33 De Vries, *Dutch Rural Economy*, 136–55.

34 Gelderblom, "Golden Age of the Dutch Republic."

35 H. Soly, "The 'Betrayal' of the Sixteenth-Century Bourgeoisie," *Acta Historiae* 8 (1975): 31–49.

36 *Political Economy of the Dutch Republic*, 174.

37 Gelderblom, "Golden Age of the Dutch Republic."

38 Ibid.; De Vries and Woude, *First Modern Economy*.

39 Theodore K. Rabb, *Enterprise and Empire: Merchant and Gentry Investment in the Expansion of England, 1575–1630* (Cambridge, MA: Harvard University Press, 1967).

40 Oscar Gelderblom, "From Antwerp to Amsterdam: The Contribution of Merchants from the Southern Netherlands to the Rise of the Amsterdam Market," *Review: A Journal of the Fernand Braudel Center* 26, no. 3 (2003), pp. 247–82.

41 William R. Scott, *The Constitution and Finance of English, Scottish and Irish Joint-stock Companies to 1729*, vol. 1 (Cambridge: Cambridge University Press,

1951), 193–97 ; Kirti N. Chaudhuri, *The English East India Company*, vol. 6, *The Emergence of International Business 1200-1800* (London: Routledge, 1999): 57, 66 and 209–20; Oscar Gelderblom and Joost Jonker, "Completing a Financial Revolution: The Finance of the Dutch East India Trade and the Rise of the Amsterdam Capital Market, 1595–1612," *Journal of Economic History* 64 no. 3 (2004): 641–72.

42 *Political Economy of the Dutch Republic*, 238.

43 Ibid., 239.

44 De Vries and Woude, *First Modern Economy*, 388.

45 Greg Clark, "The Political Foundations of Modern Economic Growth: England, 1540–1800," *Journal of Interdisciplinary History* 26 no. 4 (1996): 563–88.

46 *Political Economy of the Dutch Republic*, 245.

47 Ibid., 245.

48 Dietz, *English Public Finance*, 305–27.

49 *Political Economy of the Dutch Republic*, 245.

50 Gelderblom, "Golden Age of the Dutch Republic."

51 Ibid.

52 Ron Harris, *Industrializing English Law: Entrepreneurship and Business Organization, 1720–1844* (Cambridge: Cambridge University Press, 2000), 43.

53 *Political Economy of the Dutch Republic*, 245.

54 Oscar Gelderblom, *Cities of Commerce: The Institutional Foundation of International Trade* (Princeton, NJ: Princeton University Press, 2013).

55 Ibid.

56 Gelderblom, "From Antwerp to Amsterdam," 247–83.

57 Gelderblom, *Cities of Commerce*.

58 William R. Thompson, *Great Power Rivalries* (Columbia: University of South Carolina Press, 1999), 176.

59 Jonathan I. Israel, *Dutch Primacy in World Trade, 1585–1740* (Oxford: Clarendon Press, 1990), 101.

60 Gelderblom, "Golden Age of the Dutch Republic"; Joel Mokyr, "The Industrial Revolution and the Netherlands: Why Did It Not Happen?" Prepared for the 150th Anniversary Conference Organized by the Royal Dutch Economic Association, Amsterdam, December 10–11, 1999.

61 Gelderblom, "Golden Age of the Dutch Republic."

62 Israel, *Dutch Primacy in World Trade*, 384.

63 Ibid., 357.

64 Ibid., 358.

65 Mokyr, "The Industrial Revolution and the Netherlands."

9 Patenting: Institutionalizing Innovation

1 P. J. Federico, "Origin and Early History of Patents," *Journal of the Patent Office Society* 11 (1929): 292–305.

2 Christopher May and Susan K. Sell, *Intellectual Property Rights: A Critical History* (Boulder, CO: Lynne Rienner, 2006).

3 Nuno Pires de Carvalho, Introduction to *The TRIPS Regime of Patent Rights* (Alphen aan den Rijn: Kluwer Law International, 2010), 1–26.

4 May and Sell, *Intellectual Property Rights*.

5 Neil Davenport, *The United Kingdom Patent System: A Brief History with Bibliography* (Portsmouth: K. Mason, 1979).

6 Exceptions are secret patents, usually with regard to government secrets. These start in mid-nineteenth-century England.

7 Renée Marlin-Bennett, *Knowledge Power: Intellectual Property, Information, and Privacy* (Boulder, CO: Lynne Rienner, 2004).

8 Carvalho, Introduction, 1–26.

9 There are others, but they are incidental.

10 Elizabeth Read Foster, "The Procedure of the House of Commons against Patents and Monopolies 1621–1624," in *Law, Liberty, and Parliament: Selected Essays on the Writings of Sir Edward Coke*, ed. Allen D. Boyer (Indianapolis: Liberty Fund, 2004), 302–27.

11 Christine Macleod, *Inventing the Industrial Revolution* (Cambridge: Cambridge University Press, 1989).

12 "English Statute of Monopolies of 1623," Franklin Pierce Center for Intellectual Property, University of New Hampshire School of Law. http://ipmall.info/hosted_resources/lipa/patents/English_Statute1623.pdf.

13 Davenport, *United Kingdom Patent System*.

14 Macleod, *Inventing the Industrial Revolution*.

15 Joel Mokyr, "Entrepreneurship and the Industrial Revolution in Britain," in *The Invention of Enterprise: Entrepreneurship from Ancient Mesopotamia to Modern Times*, ed. David S. Landes, Joel Mokyr, and William J. Baumol (Princeton, NJ: Princeton University Press, 2010), 183–210.

16 Macleod, *Inventing the Industrial Revolution*.

17 Consider, for instance, England versus the Netherlands. In the seventeenth century both had a strong colonial presence and were in competition (the Netherlands could even be said to have been the stronger of the two). Both were fairly capitalistic as well. However, throughout the eighteenth century Britain rose while the Dutch empire waned. See Joel Mokyr, "The Industrial Revolution and the Netherlands: Why Did It Not Happen?" *De Economist* 148, no. 4 (2000): 503–20.

18 Macleod, *Inventing the Industrial Revolution*.

19 Christopher May, "The Hypocrisy of Forgetfulness: The Contemporary Significance of Early Innovations in Intellectual Property," *Review of International Political Economy* 14, no. 1 (2007): 1–25; Timur Kuran, "Entrepreneurship in Middle Eastern History," in *The Invention of Enterprise*, 62–87.

20 Graeme Lang, "Structural Factors in the Origins of Modern Science: A Comparison of China and Europe," in *East Asian Cultural and Historical Perspectives*, ed. Steven Totosy de Zepetnek and Jennifer W. Jay (Alberta: University of Alberta, 1997), 71–96.

21 K. N. Chadhuri, *Trade and Civilization in the Indian Ocean: An Economic History from the Rise of Islam to 1750* (Cambridge: Cambridge University Press, 1985).

22 David S. Landes, *The Wealth and Poverty of Nations: Why Some Are So Rich and Some So Poor* (London: Little, Brown, 1998).

23 Ibid.

24 Kuran, "Entrepreneurship in Middle Eastern History," 62–87.

25 Darren Acemoglu and James A. Robinson, "Economic Backwardness in Political Perspective," *American Political Science Review* 100, no. 1 (2006): 115–31.

26 For more on the Chinese attitude toward novelty and its effect on patents, see William P. Alford, "Don't Stop Thinking about … Yesterday: Why There Was No Indigenous Counterpart to Intellectual Property Law in Imperial China," *Journal of Chinese Law* 7, no. 3 (1993): 3–34. For more on the Muslim attitude toward novelty and its effect on inventiveness, see Timur Kuran, "Entrepreneurship in Middle Eastern History," 62–87. In general, most established religions can be said to be inherently against novelty. However, the important issue is how this oppositional tendency is interpreted and used at the time. In China, Confucianism was used to oppose novelty. In Muslim empires, it changes according to time and place (certain periods stress the opposition to novelty, while others ignore or find a way around it). The same is true for the various branches of Christianity.

27 Landes, *Wealth and Poverty of Nations*.

28 Alford, "Don't Stop Thinking about … Yesterday," 3–34.

29 Landes, *Wealth and Poverty of Nations*.

30 Timur Kuran, "Entrepreneurship in Middle Eastern History," 62–87.

31 Acemoglu and Robinson, "Economic Backwardness in Political Perspective," 115–31.

32 Landes, *Wealth and Poverty of Nations*.

33 Yiping Yang, "The 1990 Copyright Law of the People's Republic of China," *Pacific Basin Law Journal* 11 (1992): 260–84.

34 Frank D. Prager, "Brunelleschi's Patent," *Journal of the Patent Office Society* 28, no. 2 (1946): 109–35.

35 Ibid.

36 May and Sell, *Intellectual Property Rights*.

37 Translated from the Archives of the Florentine State, cited in Prager, "Brunelleschi's Patent," 109–35.

38 Ibid.

39 Ibid.

40 May and Sell, *Intellectual Property Rights*.

41 Prager, "Brunelleschi's Patent," 109–35.

42 May and Sell, *Intellectual Property Rights*.

43 Daron Acemoglu and James A Robinson, *Why Nations Fail: The Origins of Power, Prosperity, and Poverty* (New York: Crown Business, 2012).

44 US Congress, "Article I, Section 8,", *United States Constitution Online* (1992), accessed August 22, 2011, http://www.usconstitution.net/const.html#A1Sec8.

45 Edward C. Walterscheid, "Early Evolution of the United States Patent Law: Antecedents (5 Part I)," *Journal of the Patent and Trademark Office Society* 78 (1996): 615–39.

46 Paul Israel, *From Machine Shop to Industrial Laboratory* (Baltimore: Johns Hopkins University Press, 1992).

47 Generally, in societies that support free trade, all monopolies are considered illegitimate as they limit the openness of the market. Certain monopolies are allowed as deviations, on the basis that in their specific case they are necessary. This is the case with patents and certain government monopolies. Note that many people argue against any monopolies.

48 Israel, *From Machine Shop to Industrial Laboratory*.

49 Zorina B. Khan, *The Democratization of Invention: Patents and Copyrights in American Economic Development, 1790–1920* (Cambridge: Cambridge University Press, 2005).

50 Ibid.

51 Ibid.

52 Naomi R. Lamoreaux and Kenneth L. Sokoloff, "Intermediaries in the U.S. Market for Technology, 1870–1920," NBER Working Paper, no. 9016 (2002).

53 Khan, *Democratization of Invention*.

10 The Steam Engine and the Rise of the British Empire

1 "International Energy Outlook 2016," US Energy Information Administration, https://www.eia.gov/outlooks/ieo/electricity.cfm; Wendell H. Wiser, *Energy Resources: Occurrence, Production, Conversion, Use* (New York: Springer-Verlag, 2000). Steam turbines use steam to create a rotary motion, and thus can be used for generators, this as opposed to the push/pull motion of the steam engine. Both rely on steam. The first practical turbine was invented in 1884.

2 Jack A. Goldstone, "Efflorescences and Economic Growth in World History: Rethinking the 'Rise of the West' and the Industrial Revolution," *Journal of World History* 13, no. 2 (2002): 323–89.

3 John U. Nef, *The Rise of the British Coal Industry* (London: Routledge, 1966).

4 Reports from the early eighteenth century put it at a few thousand pounds. See Allan R. Griffin, *Mining in the East Midlands 1550–1947* (London: Frank Cass & Company, 1971).

5 Michael W. Flinn, *The History of the British Coal Industry*, vol. 2, *1700–1830* (Oxford: Oxford University Press, 1984).

6 Griffin, *Mining in the East Midlands 1550–1947*.

7 A title similar to a count or prince, used primarily in the Germanic states; Hesse was a principality in Germany.

8 L. T. C. Rolt, *Thomas Newcomen: The Prehistory of the Steam Engine* (New York: Augustus M. Kelley, 1968).

9 Considering this was the time of a civil war in Britain, it is also possible his work was disrupted by political events.

10 Rolt, *Thomas Newcomen*.

11 Ibid.

12 Jack A. Goldstone, "The Problem of the Early Modern World," *JESHO* 41, no. 3 (1998): 249–84.

13 Goldstone, "Efflorescences and Economic Growth in World History," 323–89.

14 Shannon R. Brown, "The Ewo Filature: A Study in the Transfer of Technology to China in the 19th Century," *Technology and Culture* 20, no. 3 (1979): 550–68.

15 Nathan Rosenberg, "The Impact of Technological Innovation," in *The Positive Sum Strategy: Harnessing Technology for Economic Growth*, ed. Ralph Landau and Nathan Rosenberg (Washington, DC: National Academies Press, 1986), 17–32.

16 Christine MacLeod, *Inventing the Industrial Revolution: The English Patent System, 1660–1800* (Cambridge: Cambridge University Press, 2002): 148–49.

17 Ibid., 106–9.

18 Ibid., 176–7.

19 With the exception of Papin, who neither remained in Britain nor was truly successful.

20 William Rosen, *The Most Powerful Idea in the World* (New York: Random House, 2010).

21 Gillian Sutherland, "Education," in *Social Agencies and Institutions*, ed. F. M. L. Thompson (Cambridge: Cambridge University Press, 1990), doi:10.1017/CHOL9780521257909.004.

22 H. W. Dickinson, *A Short History of the Steam Engine* (London: Frank Cass, 1963).

23 John Carroll Power, *The Rise and Progress of Sunday Schools: A Biography of Robert Raikes and William Fox* (New York: Sheldon & Company, 1863).

24 R. A. Houston, *Scottish Literacy and the Scottish Identity: Illiteracy and Society in Scotland and Northern England, 1600–1800* (Cambridge: Cambridge University Press, 2002), 3.

25 John E. Craig, "The Expansion of Education," *Review of Research in Education* 9 (1981): 151–213.

26 Dickinson, *Short History of the Steam Engine*.

27 Linda Colley, *Britons: Forging the Nation, 1688–1837* (New Haven, CT: Yale University Press, 1992).

28 Robert Hart, "Reminiscences of James Watt," *Transactions of the Glasgow Archeological Society* 1 (1859): 1–7.

29 Dickinson, *Short History of the Steam Engine*.

30 Lawrence Stone, "Literacy and Education in Britain 1640–1900," *Past & Present* 42 (1969): 69–139.

31 C. Hutton, G. Shaw and R. Pearson, eds., *The Philosophical Transactions of the Royal Society in London, from Their Commencement in 1665 to the Year 1800*, vol. 4. Abridged (London: C. & R. Baldwin, 1809); for the publication of D. Papin's experiments, see J. Lowthorp, ed., *The Philosophical Transactions and Collections, to the End of the Year 1700*, abridged (London: 1705).

32 Colley, *Britons: Forging the Nation*.

33 Dickinson, *Short History of the Steam Engine*.

34 Colley, *Britons: Forging the Nation*.

35 Neil McKendrick, "The Consumer Revolution of Eighteenth Century Britain," in *The Birth of a Consumer Society: The Commercialization of Eighteenth-Century Britain*, ed. Neil McKendrick, John Brewer and J. H. Plumb (Bloomington: Indiana University Press, 1985), 9–33.

36 Rolt, *Thomas Newcomen*.

37 Dickinson, *Short History of the Steam Engine.*
38 Johan Schot, "The Usefulness of Evolutionary Models for Explaining Innovation: the Case of the Netherlands in the Nineteenth Century," *History and Technology* 14, no. 3 (1998): 173–200.
39 Guilds were organizations that held monopolies on one trade or another (similar to modern unions). In countries that recognized guilds, one could not work in certain trades without belonging to their guild (unless they had a special dispensation from the monarch). Guilds occasionally exercised considerable political power and pressure on the government, depending on their economic power and their connections in government. They were often opposed to mechanical innovations because they felt they encroached on their livelihood. For instance, the mechanical weaving machinery of Arkwright made it possible for factory owners to hire low-wage workers rather than pay artisans for skilled work.
40 Schot, "The Usefulness of Evolutionary Models for Explaining Innovation," 173–200.
41 Ibid.
42 Ibid.
43 Fritz Redlich, "The Leaders of the German Steam-Engine Industry during the First Hundred Years," *Journal of Economic History* 2, no. 4 (1944): 121–48.
44 Justin Yifu Lin, "The Needham Puzzle: Why the Industrial Revolution Did Not Originate in China," *Economic Development and Cultural Change* 43, no. 2 (1995): 269–92.
45 Ladislao Reti, "The Double-Acting Principle in East and West," *Technology and Culture* 11, no. 2 (1970): 178–200.
46 Jack A. Goldstone, "The Rise of the West—or Not? A Revision to Socio-economic History," *Sociological Theory* 18, no. 2 (2000): 175–94.
47 Brown, "The Ewo Filature," 550–68.
48 Lin, "The Needham Puzzle," 269–92.
49 Goldstone, "The Rise of the West—or Not?" 175–94.
50 Joseph Needham, "Science and Society in East and West," *Science and Society* 28, no. 4 (1964): 385–408.
51 Shannon R. Brown, "The Transfer of Technology to China in the Nineteenth Century: The Role of Direct Foreign Investment," *Journal of Economic History* 39, no. 1 (1979): 181–97.
52 Rolt, *Thomas Newcomen.*
53 John Bach McMaster, *A History of the People of the United States: From the Revolution to the Civil War,* vol. 1 (New York: Cosimo Inc., 2005).
54 Hans Binswanger, "Agricultural Mechanization: A Comparative Historical Perspective," *World Bank Research Observer* 1 (1986): 45.

11 American Mass Production and the Rise of the United States

1 Stephen H. Haber, "Industrial Concentration and the Capital Markets: A Comparative Study of Brazil, Mexico, and the United States, 1830–1930,"

Journal of Economic History 51 (1991): 559–80; S. Haber and H. S. Klein, "The Economic Consequences of Brazilian Independence," in *How Latin America Fell Behind: Essays on the Economic Histories of Brazil and Mexico, 1800–1914*, ed. Stephen H. Haber (Palo Alto, CA: Stanford University Press, 1997).

2 It fell from $738 to $700—Haber and Klein, "The Economic Consequences of Brazilian Independence"; Robert B. Ekelund, Fernando C. Zanella and David N. Laband, "Monarchy, Monopoly and Mercantilism: Brazil Versus the United States in the 1800s," *Public Choice* 116 (September 2003): 392.

3 Charles R. Morris, *The Dawn of Innovation: The First American Industrial Revolution* (New York: Public Affairs, 2012).

4 Ibid.

5 Xiuzhen Janice Li, "Standardisation, Labour Organisation and the Bronze Weapons of the Qin Terracotta Warriors" (PhD diss., University College London, 2012).

6 Carolyn C. Cooper, "The Portsmouth System of Manufacture," *Technology and Culture* 25 (1984); Daron Acemoglu and James Robinson, *Why Nations Fail: The Origins of Power, Prosperity, and Poverty* (New York: Crown, 2012), 32.

7 Morris, *Dawn of Innovation*.

8 Hans Binswanger, "Agricultural Mechanization: A Comparative Historical Perspective," *World Bank Research Observer* 1 (1986): 45.

9 David J. Jeremy, "Innovation in American Textile Technology during the Early 19th Century," *Technology and Culture* 14 (1973): 40–76.

10 Rights of women and blacks kept evolving over time. However, in general, and especially in the North, these were relatively much more progressive than in Brazil and Mexico, and more progressive than in many countries of Europe. Similarly, rights of immigrants from Africa and Asia kept evolving over time. They became full equal only after 1965. Yet, immigration was much more open to people of different countries than in Mexico and Brazil.

11 US Congress, "The Charters of Freedom—A New World is at Hand," *The Declaration of Independence: A Transcription* (1776), accessed July 16, 2014, http://www.archives.gov/exhibits/charters/declaration_transcript.html.

12 Acemoglu and Robinson, *Why Nations Fail*.

13 Jane O'Brien, " 'Proof' Jamestown Settlers Turned to Cannibalism," *BBC News* May 1, 2013.

14 Acemoglu and Robinson, *Why Nations Fail*.

15 Ekelund, Zanella and Laband, "Monarchy, Monopoly and Mercantilism," 385.

16 Acemoglu and Robinson, *Why Nations Fail*.

17 "Northwest Ordinance," West's Encyclopedia of American Law, accessed July 17, 2014, http://www.encyclopedia.com/doc/1G2-3437704859.html.

18 Sheryll Patterson-Black, "Women Homesteaders on the Great Plains Frontier," *Frontiers: A Journal of Women Studies* 1 (Spring, 1976): 70–74.

19 T. Williams, "The Homestead Act: Our Earliest National Asset Policy," Working paper 3–5, St. Louis, Washington University, 2003.

20 Paul P. Christensen, "Land Abundance and Cheap Horsepower in the Mechanization of the Antebellum United States Economy," *Explorations in Economic History* 18 (1981): 309–29.

21 Patterson-Black, "Women Homesteaders on the Great Plains Frontier," 70–74.

22 Williams, *Homestead Act.*

23 Patterson-Black, "Women Homesteaders on the Great Plains Frontier," 70–74.

24 David Igler, *Industrial Cowboys: Miller & Lux and the Transformation of the Far West, 1850–1920* (Berkeley: University of California Press, 2001).

25 Micahel Magliari, "Free Soil, Unfree Labor: Cave Johnson Couts and the Binding of Indian Workers in California, 1850–1867," *Pacific Historical Review* 73, no. 3 (2004): 349–90.

26 Morris, *Dawn of Innovation.*

27 Williams, *Homestead Act.*

28 Trina R. Williams Shanks, "The Homestead Act: A Major Asset-Building Policy in American History," in *Inclusion in the American Dream: Assets, Poverty, and Public Policy*, ed. M.W. Sherraden (New York: Oxford University Press, 2005), 20–41.

29 John H. Coatsworth, "Obstacles to Economic Growth in Nineteenth-Century Mexico," *American Historical Review* 83 (February 1978): 88.

30 Acemoglu and Robinson, *Why Nations Fail.*

31 Ibid.

32 Edward Beatty, "Approaches to Technology Transfer in History and the Case of Nineteenth-Century Mexico," *Comparative Technology Transfer and Society* 1 (August 2003): 167–97.

33 Simon Miller, "Mexican Junkers and Capitalist Haciendas, 1810–1910: The Arable Estate and the Transition to Capitalism between the Insurgency and the Revolution," *Journal of Latin American Studies* 22 (May 1990): 261.

34 François Bourguignon and Christian Morrisson, "Inequality Among World Citizens: 1820–1992," *American Economic Review* 92 (September 2002): 727–44.

35 Jeniffer Brown, "Ejidos and Comunidades in Oaxaca, Mexico: Impact of the 1992 Reforms," *RDI Reports on Foreign Aid and Development*, no. 120, February 2004, 1–32.

36 Cámara de Diputados, "Articale 27," Constitución Política de los Estados Unidos Mexicanos (1917), accessed August 13, 2014, http://www.diputados.gob.mx/leyesbiblio/htm/1.htm.

37 Brown, "Ejidos and Comunidades in Oaxaca, Mexico."

38 Cámara de Diputados, "Articale 27."

39 Haber and Klein, "The Economic Consequences of Brazilian Independence"; Ekelund, Zanella and Laband, "Monarchy, Monopoly and Mercantilism," 392.

40 S. Graeff-Hönninger, Martin Gauder and W. Claupein, "The Impact of a Growing Bioethanol Industry on Food Production in Brazil," *Applied Energy* 88 (March 2011): 672–79.

41 Márcia Maria Menendes Motta, "Justice and Violence in the Lands of the Assecas (Rio De Janeiro, 1729–1745)," *Historia Agraria: Revista de agricultura e historia* 58 (December 2012): 13–37.

42 Graeff-Hönninger, Gauder and Claupein, "The Impact of a Growing Bioethanol Industry on Food Production in Brazil," 672–79.

43 Zanella, Ekelund and Laband, "Monarchy, Monopoly and Mercantilism," 381–98.

44 Nathaniel H. Leff, "Economic Development in Brazil, 1811–1913," in *How Latin America Fell Behind*, 34–64.

45 E. Harris, B. Mueller and L. J. Alston, "De Facto and De Jure Property Rights: Land Settlement and Land Conflict on the Australian, Brazilian and U.S.," in NBER Working Paper Series, 2009.

46 Ibid.

47 Ibid.

48 Stephen H. Haber, "Financial Markets and Industrial Development: A Comparative Study of Governmental Regulation, Financial Innovation and Industrial Structure in Brazil and Mexico, 1840–1940," in *How Latin America Fell Behind*, 146–54.

49 Chaim M. Rosenberg, *The Life and Times of Francis Cabot Lowell, 1775–1817* (Lanham, MD: Lexington Books, 2010).

50 Ibid.

51 Morris, *Dawn of Innovation*.

52 Ibid.

53 Ibid.

54 Ibid.

55 Jeremy, "Innovation in American Textile Technology during the Early 19th Century," 40–76.

56 Robert P. Sutton, "Sectionalism and Social Structure: A Case Study of Jeffersonian Democracy," *Virginia Magazine of History and Biography*, January 1972, 70–84.

57 Jeremy, "Innovation in American Textile Technology during the Early 19th Century," 40–76.

58 Morris, *Dawn of Innovation*.

59 Ibid.

60 Ibid.

61 Ibid.

62 Ibid.

63 Acemoglu and Robinson, *Why Nations Fail.*

64 Haber, "Industrial Concentration and the Capital Markets," 559–80.

65 Morris, *Dawn of Innovation*.

66 Haber, "Industrial Concentration and the Capital Markets," 559–80.

67 Acemoglu and Robinson, *Why Nations Fail.*

68 Ibid.

69 Federal Trade Commission, Protecting America's Consumers, http://www.ftc.gov/about-ftc/bureaus-offices/bureau-competition/about-bureau-competition.

70 US Department of Justice, "History of the Antitrust Division," http://www.justice.gov/atr/about/division-history.html.

71 G. A. Field, "The Status Float Phenomenon," *Business Horizons* (1970): 45–52.

72 Morris, *Dawn of Innovation*.

73 Haber, "Industrial Concentration and the Capital Markets," 559–80.

74 Ibid., 559–80.

75 Beatty, "Approaches to Technology Transfer in History," 167–97.

76 Jürgen Buchenau, "Small Numbers, Great Impact: Mexico and Its Immigrants, 1821–1973," *Journal of American Ethnic History* 20 (2001): 23–50.

77 Haber, "Industrial Concentration and the Capital Markets," 559–80.

78 Zanella, Ekelund and Laband, "Monarchy, Monopoly and Mercantilism," 381–98.

79 Harris, Mueller and Alston, "De Facto and De Jure Property Rights."

80 Zanella, Ekelund and Laband, "Monarchy, Monopoly and Mercantilism," 381–98.

81 Haber, "Industrial Concentration and the Capital Markets," 559–80.

82 "Mass Production," *Dictionary of American History*, accessed June 19, 2014, http://www.encyclopedia.com.

83 Giorgio Riello, "Boundless Competition: Subcontracting and the London Economy in the Late Nineteenth Century Cover," *Enterprise & Society*, 13 (September 2012): 504–37.

84 Besides these two, US immigration policy also suffered other constraints. Starting with the War of Independence, citizenship was restricted to free white men. That was extended to African Americans after the end of the Civil War and to Asians only after 1945. Until 1965, there were unequal quotas of immigrants from various countries. After 1965, the same fixed number applied to all countries. The point being made in this chapter is that relative to Mexico and Brazil, US immigration was far more open, equal and empowering.

85 Probably first alluded to in J. De Crevecoeur and Hector St. John, *Letters from An American Farmer* (1782).

86 US Bureau of the Census, *Historical Statistics of the United States, Colonial Times to 1970*, vol. 2 (Washington, DC: Bicentennial Edition, 1975).

87 Angus Maddison, *Historical Statistics on World Economy 1–200*, accessed July 3, 2017, https://www.google.com/search?q=maddison+worl+population+20+c ountries&oq=maddison+worl+population+20+countries&aqs=chrome..69i57 .8599j0j7&sourceid=chrome&ie=UTF-8.

88 Simon Kuznets, "Two Centuries of Economic Growth: Reflections on U.S. Experience," *American Economic Review* 67, no. 1 (February 1977): 1–14.

89 Michael C. LeMay, *From Open Door to Dutch Door: An Analysis of U.S. Immigration Policy since 1820* (Westport, CT; London: Praeger, 1987).

90 Williams, *Homestead Act*.

91 Stephen Castles and Mark J. Miller, *The Age of Migration* (New York: Guilford Press, 2003); Joseph P. Ferrie, "The Entry into the U.S. Labor Market of Antebellum European Immigrants, 1840–1860," *Explorations in Economic History* 34 (1997): 295–330.

92 Castles and Miller, *Age of Migration*.

93 LeMay, *From Open Door to Dutch Door*.

94 Zanella, Ekelund and Laband, "Monarchy, Monopoly and Mercantilism," 381–98.

95 Nancy S. Landale and Avery M. Guest, "Generation, Ethnicity, and Occupational Opportunity in Late 19th Century America," *American Sociological Review* 55 (April 1990): 280–96.

96 Peter J. Hill, "Relative Skill and Income Levels of Native and Foreign Born Workers in the United States," *Explorations in Economic History* 12 (January, 1975): 47–60.

97 Castles and Miller, *Age of Migration*.

98 Patricia Cloud and David W. Galenson, "Chinese Immigration and Contract Labor in the Late Nineteenth Century," *Explorations in Economic History* 24 (1987): 22–42.

99 Magliari, "Free Soil, Unfree Labor," 349–90.

100 Henry K. Norton, *The Story of California from the Earliest Days to the Present* (Chicago: A. C. McClurg, 1924), 283–96.

101 Andrew Gyory, *Andrew Gyory* (Chapel Hill: University of North Carolina Press, 1998).

102 Landale and Guest, "Generation, Ethnicity, and Occupational Opportunity," 280–96; Kuznets, "Two Centuries of Economic Growth," 1–14.

103 Riello, "Boundless Competition," 504–37.

104 Morris, *Dawn of Innovation*.

105 Ibid.

106 Ibid.

107 Ibid.

108 Ross Thomson, *Structures of Change in the Mechanical Age: Technological Innovation in the United States, 1790–1865* (Baltimore: Johns Hopkins University Press, 2009).

109 John Majewski, "Who Financed the Transportation Revolution? Regional Divergence and Internal Improvements in Antebellum Pennsylvania and Virginia," *Journal of Economic History* 56 (December 1996): 763–88.

110 Lawrence H. Summers and Daniel M.G. Raff, "Did Henry Ford Pay Efficiency Wages?" *Journal of Labor Economics* 5 (October 1987): 57–86.

111 Ibid.

112 T. G. Powell, "Priests and Peasants in Central Mexico: Social Conflict During 'La Reforma'," *Hispanic American Historical Review* 57 (May 1977): 296–313.

113 Laura Valeria González-Murphy, *Protecting Immigrant Rights in Mexico: Understanding the State-civil Society Nexus* (New York: Routledge).

114 Buchenau, "Small Numbers, Great Impact," 23–50.

115 Robert McCaa, "The Peopling of Mexico from Origins to Revolution," in *The Population History of North America*, ed. Richard Steckel and Michael Haines (Cambridge: Cambridge University Press, 2000), 241–304.

116 Luis Bertola and Jeffrey G. Williamson, "Globalization in Latin America before 1940," NBER Working Paper, no. w9687, May 2003, 17.

117 Buchenau, "Small Numbers, Great Impact," 23–50.

118 Ibid.

119 Ibid.

120 Ibid.

121 Ibid.

122 Ibid.

123 The United States abolished slavery after the Civil War in 1865. Even before then, slavery occurred in only the Southern states. Most of the innovation

and industrial innovation took place in the Northern and Western states with no slavery.

124 Haber and Klein, "The Economic Consequences of Brazilian Independence."

125 Harris, Mueller and Alston, "De Facto and De Jure Property Rights."

126 Ibid.

127 Zanella, Ekelund and Laband, "Monarchy, Monopoly and Mercantilism," 381–98.

128 Ibid.

129 Thomas Skidmore, "Racial Ideas and Social Policy in Brazil, 1870–1940," in *The Idea of Race in Latin America, 1870–1940*, ed. R. Graham (Austin: Texas University Press, 1990), 7–36.

130 Magnus Mörner and Harold Sims, *Adventurers and Proletarians: The Story of Migrants in Latin America* (Pittsburgh: University of Pittsburgh Press, 1985).

131 IMF (International Monetary Fund), *Report for Selected Country Groups and Subjects (PPP) Valuation of Country GDP* (Washington, DC: International Monetary Fund, 2017).

132 "5 Facts about the U.S. Rank in Worldwide Migration," Pew Research Center, http://www.pewresearch.org/fact-tank/2016/05/18/5-facts-about-the-u-s-rank-in-worldwide-migration/.

133 Ibid.

134 R. T. Herman and Robert L. Smith, *Immigrant, Inc.: Why Immigrant Entrepreneurs are Driving the New Economy* (Hoboken, NJ: John Wiley & Sons, 2010).

135 Steven Ballmer, *The New American Fortune 500*, Partnership for a New America, June 2011, accessed March 6, 2018, http://www.newamericaneconomy.org/sites/all/themes/pnae/img/new-american-fortune-500-june-2011.pdf.

136 Stuart Anderson, "Immigrants and Billion Dollars Startups," National Foundation for American Policy, March 2016, accessed September 17, 2017, http://nfap.com/wp-content/uploads/2016/03/Immigrants-and-Billion-Dollar-Startups.NFAP-Policy-Brief.March-2016.pdf.

137 AnnaLee Saxenian, *Silicon Valley's New Immigrant Entrepreneurs* (San Francisco: Public Policy Institute of California, 1999); Gary Hamel, "Bringing Silicon Valley Inside," *Harvard Business Review* (September-October, 1999), accessed March 8, 2018, https://hbr.org/1999/09/bringing-silicon-valley-inside.

138 Vivek Wadhwa, AnnaLee Saxenian and Francis Daniel Siciliano II, "Then and Now: America's New Immigrant Entrepreneurs," Ewing Marion Kauffman Foundation Research Paper (2012). Posted October 10, 2012, accessed March 6, 2018, https://papers.ssrn.com/sol3/papers.cfm?abstract_id=2159875.

139 Stav Rosenzweig, Amir Grinstein and Elie Ofek, "Social Network Utilization and the Impact of Academic Research in Marketing," *International Journal of Research in Marketing* 33, no. 4 (2016): 818–39; Peter Vandor and Nikolaus Franke, "Why Are Immigrants More Entrepreneurial?" *Harvard Business Review*, October 27, 2016, accessed March 6, 2018, https://hbr.org/2016/10/why-are-immigrants-more-entrepreneurial

140 "Competitiveness Rankings," World Economic Forum, 2016, accessed May 17, 2017, http://reports.weforum.org/global-competitiveness-report-2015–2016/competitiveness-rankings/

141 One example is Israeli food; see Gabriel Davidovich-Weisberg, "Want Cheaper Israeli Food? Best Buy It in the United States," *Haaretz*, January 12, 2014, accessed May 26, 2017, http://www.haaretz.com/israel-news/business/.premium-1.568169.

142 Saxenian, *Silicon Valley's New Immigrant Entrepreneurs*; AnnaLee Saxenian and Yasuyuki Motoyama, *Local and Global Networks of Immigrant Professional in Silicon Valley* (San Francisco: Public Policy Institute of California, 2002).

12 Lessons

1 Gerard J. Tellis, *Unrelenting Innovation: How to Create a Culture for Market Dominance* (San Francisco: Jossey-Bass, 2013).

2 Daron Acemoglu and James A. Robinson, *Why Nations Fail: The Origin of Power, Prosperity, and Poverty* (New York: Crown Business, 2012).

3 Tellis, *Unrelenting Innovation*.

4 Ibid.

5 Rajesh Chandy and Gerard J. Tellis, "The Incumbent's Curse: Incumbency, Size, and Radical Innovation," *Journal of Marketing* 64 (July 2000): 1–17.

6 Stav Rosenzweig, "Non-Customers as Initiators of Radical Innovation," *Industrial Marketing Management* 66, no. 6 (2017): 1–12.

7 Robyn Meredith, *The Elephant and the Dragon: The Rise of India and China and What It Means for All of Us* (New York: W. W. Norton, 2007); William M. Overholt, *The Rise of China: How Economic Reform Is Creating a New Superpower* (New York: W. W. Norton, 1993).

Index